Serious Leisure and Nature

Also by Robert A. Stebbins

THE ORGANIZATIONAL BASIS OF LEISURE PARTICIPATION: A Motivational Exploration

BETWEEN WORK AND LEISURE: The Common Ground of Two Separate Worlds

CHALLENGING MOUNTAIN NATURE: Risk, Motive, and Lifestyle in Three Hobbyist Sports

SERIOUS LEISURE: A Perspective of Our Time

THE PIVOTAL ROLE OF LEISURE EDUCATION: Finding Personal-Fulfillment in this Century (*co-edited*)

PERSONAL DECISIONS IN THE PUBLIC SPQUARE: Beyond Problem Solving into a Positive Sociology

LEISURE AND CONSUMPTION: Common Ground, Separate Worlds

Serious Leisure and Nature

Sustainable Consumption in the Outdoors

Lee Davidson
Victoria University of Wellington, New Zealand

Robert A. Stebbins
University of Calgary, Canada

First published 2011 by
PALGRAVE MACMILLAN

Palgrave Macmillan in the UK is an imprint of Macmillan Publishers Limited,
registered in England, company number 785998, of Houndmills, Basingstoke,
Hampshire RG21 6XS.

Palgrave Macmillan in the US is a division of St Martin's Press LLC,
175 Fifth Avenue, New York, NY 10010.

Palgrave Macmillan is the global academic imprint of the above companies
and has companies and representatives throughout the world.

Palgrave® and Macmillan® are registered trademarks in the United States,
the United Kingdom, Europe and other countries.

ISBN 978-1-349-32168-1   ISBN 978-0-230-29937-5 (eBook)
DOI 10.1057/9780230299375

A catalogue record for this book is available from the British Library.

A catalog record for this book is available from the Library of Congress.

10   9   8   7   6   5   4   3   2   1
20   19   18   17   16   15   14   13   12   11

To Elery Hamilton-Smith

Contents

Preface

This book examines, from the angles of environmental sustainability and leisure consumption, a modern leisure phenomenon which we call nature challenge activity (NCA), a distinctive type of outdoor pursuit that, in one form or another, appeals to all ages. The NCA is leisure whose core activity or activities centre on meeting a natural test posed by one or more of six elements: (1) air, (2) water, (3) land, (4) animals (including birds and fish), (5) plants and (6) ice or snow (sometimes both). A main reason for engaging in a particular NCA is to experience participation in its core activities pursued in a natural setting. In other words, while executing these activities, the special (aesthetic) appeal of the natural environment in which this process occurs simultaneously sets the challenge the participant seeks. At the same time many participants tend to consume the goods and services related to their NCAs in ways that are environmental friendly. Of course, as we will point out in many places in this book, other important reasons also exist for such participation, as suggested by the geographic, economic and social dimensions of the use of nature. Moreover some NCAs even have a counterpart in a line of work (e.g., professional sport fishermen, mountain guides, nature photographers), which because they constitute a livelihood will not be considered here.

Outings in nature considered as leisure activity constitute a main way in which people in the West of today use their free time. For the purposes of this book we regard nature as any natural setting *perceived* by users as at most only minimally modified by human beings. In its most general manifestation, nature thus defined is composed of one or more of the aforementioned six elements. The perceptual qualification just made is important, since nature lovers may feel they are in nature that, for example, is nonetheless imperceptibly polluted. In fact all six elements risk being sullied by this unwelcome process. Or nature lovers might recognize that pollution exists, but discount it as insignificant or unavoidable. For instance, certain kinds of trees might be dying, precipitated by an infestation of an insect, but a hike or cross-country ski through this region is still enjoyable, in part because the hikers or skiers are enthralled with other aspects of nature viewed by them as pristine (e.g., the snow, forest, geography and clean air).

Westerners have unequal access to nature, with, it would appear, those living in cities being, as a group, the most deprived in this respect. Much of city life takes place in substantially, if not entirely, artificial surroundings. Subways, buildings (notwithstanding the occasional plant or water fountain inside), streets and sidewalks (even those lined with trees), bridges and the like are the antithesis of nature as just defined it. Furthermore city parks, walk ways along rivers, public gardens and similar developments are only marginally less artificial. And the built environment is as prominent in small towns and even on farms to the extent that the latter abut one another to form a contiguous stretch of land modified by humans. This is evident in roads, buildings, planted crops and fenced pasture.

In all these settings nature has been, for the most part, far more than minimally modified for human use. This is the environmental dimension of NCA. And, for the inhabitants of these settings, nature as defined here must usually be sought elsewhere. But the question of how far they must go to find such nature constitutes another Western, in this instance, geographic dimension. In certain areas of Europe, compared with parts of Australia, New Zealand and North America, nature is rather difficult to find. People in some rural areas need drive only a few miles to find the nature they like, whereas those in many cities have to travel several hours for such experience. Moreover nature may be costly – the economic dimension – ranging from a dollar or two to enter a state or provincial park to hundreds of dollars for a seasonal pass to an alpine ski resort. By no means everyone can afford the latter.

It will also become apparent in this book that the six elements of nature are not equally appealing to all people. This is the aesthetic dimension. That is most of them are attracted only to some of the elements of nature and only to some of the NCAs available there. And there are people who care little or none at all for nature; they relish selected artificial aspects of their city, town or farm. Furthermore, there is a social dimension to the free-time use of nature. In some of this use involvement with other people is essential, particularly in the various interpersonal competitions. Yet, in many activities in nature, direct involvement with others is either optional (e.g., one may hike, photograph flowers or go orienteering alone or in a group) or impossible (e.g., alpine skiing, hang gliding, mountain biking).

The following list further describes the NCA. Always the underlying defining principle is that the setting in which it occurs is perceived as either fully natural or only minimally modified to facilitate the complex activity in question. Thus

- many NCAs are pursued using special equipment (e.g., shoes, skis, boats, skates, bicycles, rope, maps, parachutes, compasses, binoculars, hunting rifles, fishing tackle, scuba equipment, wind/surf boards, snowboards, caving lanterns);
- some NCAs are pursued in natural settings, even while, to facilitate the activity, they require certain relatively unobtrusive modifications to those settings. Examples of such modification include groomed cross-country ski trails, foot bridges and drainage ditches on hiking trails, duck hunting blinds and climbing pins driven into rock;
- some NCAs are pursued without aid of mechanical devices such as motors, which in any case, modify substantially the cherished experience of the immediate natural environment. Nevertheless some nature challenge participants use a motorboat to reach a fishing site, a sport utility vehicle to reach a hunting area, a helicopter to reach deep snow for backcountry skiing, a boat to get to a scuba site, an airplane to enable parachuting, a car to reach mushrooming terrain and so on;
- other NCAs *are* motorized, among them, water-skiing, aeronautics, snowmobiling, jet skiing, power boat racing, and beach auto and motorcycle racing. Here, for some participants, the appeal of the experience of being in nature is diluted to some extent by the noise of the machine. Others, however, interpret this noise favourably as part of the overall experience;
- by dint of the effort required to meet a particular challenge, all NCAs are serious leisure (discussed in Chapter 1). They are pursued in all three of its types: amateur, hobbyist and career volunteer;
- some NCAs may foster interpersonal competition. That is, some or all of the time, they are pursued as a sport. The attraction of the central activity done in nature remains important here, but competing against other people becomes another main reason for engaging in it.

Still not every leisure activity undertaken in nature conforms to the definition of an NCA. Activities in which participants are not, themselves, in the focal natural setting are excluded. Such is the situation for those who fly kites or model airplanes or run model boats or cars. It is the kite, plane, boat or car that experiences, as it were, the natural element of air, water or land, not the person controlling it from a distance. Furthermore, high-risk, or 'extreme' activities (defined as such by participants) are not NCAs, primarily because they shift attention away from the inherent appeal of the core activity in its natural setting to avoiding the substantial possibility of serious injury or death as well as to other distracting considerations.

We have also excluded activities that offer challenges through one of the six elements, but that are undertaken in substantially artificial circumstances. These include pursuits requiring use of swimming pools, ice rinks (indoor and outdoor), artificial climbing walls, fish and game farms, snowboard parks, velodromes, luge and bobsleigh runs, ski jumps and groomed alpine ski and snowboard runs, race tracks (e.g., auto, horse, motorcycle), courses (e.g., track and field, sled dog, model trains and cars), rodeo arenas, pastures (for raising horses, sheep, etc.), golf courses and equestrian grounds.[1] Furthermore such activities are environmentally unfriendly and, hence, unsustainable.

We are not splitting hairs in our attempt to distinguish NCAs from other outdoor leisure activities. NCAs stand out as separate and important for at least five reasons. First, it appears that many people pursue NCAs considered as a category, though we do lack precise comparative data with which to substantiate this claim. Second, these activities, unlike the others, are exclusively serious leisure, experienced in this instance in highly appealing natural settings. Its principal reward is personally fulfilling execution of a core activity done in an environment that is itself awe-inspiring. Third, in part because nature is awe-inspiring for nature challenge enthusiasts and in part because it must remain as pristine as possible for them, they often make fine champions of sustainability and the consumption of goods and services enabling the activity while causing minimal environmental damage.

Fourth, many NCAs, when pursued regularly, also contribute to their participants' physical fitness, suggesting that, for them, achieving fitness in such settings may be more appealing than doing so in artificial circumstances. Fifth, the NCAs offer greater scope for human agency than many other kinds of leisure activity. Thus, swimmers in lakes and rivers with no officially established swimming areas have the freedom to swim where they wish compared with their counterparts constrained by official boundaries in such places or limited by the edges of artificial pools. Hikers and cross-country skiers may explore off-trail, hang-gliders are free to search wind currents and chart their course of flight at will, and hobbyist sailors have an entire ocean or lake to explore (necessarily mindful of dangerous areas and changing weather patterns). Artificial areas constrain such tendencies, while scope for human agency, say participants in many of the activities considered in this book, is a central attraction of their chosen NCAs.

Nature challenge activity as a concept is a newcomer to the literatures of leisure studies, sustainability and consumption. Stebbins coined and defined the term to capture how his sample of kayakers, snowboarders and mountain climbers viewed the core activities of their hobbies. It was also clear from his study that there are many more NCAs than these three and, following further analysis of the data, that the three types of hobbyists pursue such activities in their natural settings because of the wonder the settings hold for them. More recently he has examined the complex relationship between NCAs in general and the consumption of goods and services.

This book explores systematically the idea that an NCA is pursued in the awe-inspiring natural environment of that activity. Being in awe, or wonder, of an aesthetically attractive part of nature brings a psychological dimension to our study of NCAs.

The theoretic framework guiding this book – the serious leisure perspective – roots in the concepts of leisure activity, core activity and the experience of pursuing both. Since the idea of NCA is of recent origin, scientific literature on it is in general exceedingly slim. Hence the study reported here is, of necessity, exploratory. It is designed to generate grounded theory on as wide range of NCAs as we could identify and find information on. The data are almost exclusively of the 'library' variety: where available we examined past research on particular NCAs as well as drew on biographic and autobiographic materials reporting personal involvements in them. In addition to the limited scholarly sources of journal articles, book chapters and research monographs, we looked primarily in books, videos, newspapers, magazines, newsletters and in such online sources as websites and Web 2.0 (Facebook, My Space, YouTube, etc.).

Our goal was to read about the basic features of and personal experiences in each NCA, whether old or new, pursued in the six elements, always with an eye to learning about the distinctive properties of particular nature challenges as well as their distinctive awe-inspiring qualities. In the main our analysis proceeded along lines of the six previously mentioned dimensions (environmental, geographic, economic, aesthetic, social, psychological). Additionally, and where possible, we always tried to extract from these readings how the NCA in question met the distinguishing qualities of serious leisure. It is imperative in studying serious leisure to determine that the core activities being examined are neither casual nor project-based leisure. In the end we were able to gather workable amounts (enough for generating valid inductive generalizations) of scholarly and popular material on several hundred NCAs and types of activities. Each activity will be described and examined as thoroughly as possible along the aforementioned lines.

Note

1. One might question listing groomed alpine skiing and snowboarding here. Our justification for doing so is presented in Chapter 1, n. 1.

Acknowledgements

We want to thank Philippa Grand for taking the time during a hectic conference of the American Sociological Association in 2009 to talk at length to the second author about our proposal for this book. Authors and editors dealing with scholarly books often don't meet face-to-face at all, and when they do, the encounter is commonly rushed and subject to interruptions. Nevertheless she liked our proposal, acted on it quickly and efficiently, and by mid-fall that year, we got the green light to proceed with our project.

Once writing began, the editorial ball was passed to Olivia Middleton, who promptly answered our queries about such matters as style and deadlines. Joeljones Alexander oversaw the copyediting, proof-reading and indexing phases of production. As with the first two, his work was cheerful, accurate and punctual. In short, preparation of this book has been blessed with a superb editorial team, reminding us still again how important this complicated, but unheralded, function is in publishing.

As for the writing itself we were largely on our own, with but one exception. We want to thank Chris Marshall of the Lazy Seals Freediving Club in Wellington for his help on the freediving section of Chapter 3.

1
Nature Challenge Activities

Outings in nature considered as leisure activity constitute a main way in which people in the West of today use their free time. We define nature, for the purposes of this book, as any natural setting perceived by users as at most only minimally modified by human beings. In its most general manifestation, nature is composed of one or more of the following six elements: (1) air, (2) water, (3) land, (4) animals (including birds and fish), (5) plants and (6) ice or snow (or both). The perceptual qualification just made is important, since nature lovers may feel they are in nature that, for example, is nonetheless imperceptibly polluted. Indeed all six elements risk being sullied by this offensive process. Or nature lovers might recognize that pollution exists, but discount it as insignificant. Thus, certain kinds of trees could be dying, precipitated by an infestation of an insect, but a hike or cross-country ski through this region is still enjoyable, in part because the hikers or skiers are enthralled with other aspects of nature viewed by them as pristine (e.g., the snow, forest, geography and clean air).

People in the West have unequal access to nature, with it would appear, those living in cities being, as a group, the most deprived in this regard. Much of city life takes place in substantially, if not entirely, artificial surroundings. Subways, buildings (notwithstanding the occasional plant or water fountain inside), streets and sidewalks (even those lined with trees), bridges and the like are the antithesis of nature as just defined it. Furthermore city parks, walk ways along rivers, public gardens and similar developments are only marginally less artificial. And the built environment is as prominent in small towns and even on farms, to the extent that the latter abut one another to form a contiguous stretch of land modified by humans, as evidenced in roads, buildings, planted crops and fenced pasture.

In all these settings nature has been, for the most part, far more than minimally modified for human use. This is the environmental dimension of nature challenge activity. And, for the inhabitants of these settings, nature as defined here must usually be sought elsewhere. But the question of how far they must go to find such nature constitutes another Western, in this instance, geographic dimension. In certain areas of Europe, compared with parts of Australia, New Zealand and North America, nature is quite difficult to find. People in some rural areas need drive only a few miles to find the nature they like, whereas those in many cities have to travel for several hours to find it. Moreover being in nature may be costly – the economic dimension – ranging from a dollar or two to enter a state or provincial park to hundreds of dollars for a seasonal pass to an alpine ski resort. By no means can everyone afford the latter.

It will also become apparent in this book that the six elements of nature are not equally appealing to all people (the aesthetic dimension). That is most of them are attracted only to some of the elements of nature and only to some of the activities available there. And there are people who care little or none at all for nature; they relish various artificial aspects of their city, town or farm. Furthermore, there is a social dimension to the free-time use of nature. In some of this use involvement with other people is essential, particularly in the various interpersonal competitions. Yet, in many activities in nature, direct involvement with others is either optional (e.g., one may hike, photograph flowers or go orienteering alone or in a group) or impossible (e.g., alpine skiing, hang gliding, mountain biking).

Nature challenge activity

A nature challenge activity (NCA) is a leisure pursuit whose core activity or activities centre on meeting a natural test posed by one or more of the six elements. A main reason for engaging in a particular NCA is to experience participation in its core activities pursued in a natural setting. In other words, while executing this activity, the special (aesthetic) appeal of the natural environment in which this process occurs simultaneously sets the challenge the participant seeks. But, as we will point out in many places in this book, other important reasons also exist for such participation, as suggested by the geographic, economic and social dimensions of the use of nature. Finally note that some NCAs have a counterpart in a line of work (e.g., professional sport fishermen, mountain guides, nature photographers), which because they constitute a livelihood – usually a 'devotee occupation' (Stebbins, 2004a) – fall beyond the scope of this book about leisure.

We turn now to a more detailed look at the distinctive characteristics of NCAs. Always the underlying defining principle is that the setting in which they occur is perceived as either fully natural or only minimally modified to facilitate the complex activity in question. Thus

- many NCAs are pursued using special equipment (e.g., shoes, skis, boats, skates, bicycles, rope, maps, parachutes, compasses, binoculars, hunting rifles, fishing tackle, scuba equipment, wind/surf boards, snowboards, caving lanterns);
- some NCAs are pursued in natural settings, even while, to facilitate the activity, they require certain relatively unobtrusive modifications to those settings. Examples of such modification include groomed cross-country ski trails, foot bridges and drainage ditches on hiking trails, duck hunting blinds and climbing pins driven into rock;
- some NCAs are pursued without aid of mechanical devices such as motors, which in any case modify substantially the cherished experience of the immediate natural environment. Nevertheless some nature challenge participants use a motorboat to reach a fishing site, a sport utility vehicle to reach a hunting area, a helicopter to reach deep snow for backcountry skiing, a boat to get to a scuba site, an airplane to enable parachuting, a car to reach mushrooming terrain and so on;
- other NCAs *are* motorized, among them, water-skiing, aeronautics, snowmobiling, jet skiing, power boat racing, and beach auto and motorcycle racing. Here, for some participants, the appeal of the experience of being in nature is diluted to some extent by the noise of the machine. Others, however, interpret this noise favourably as part of the overall experience;
- by dint of the effort required to meet a particular challenge, all NCAs are serious leisure (see next section). They are pursued in all three of its types: amateur, hobbyist and career volunteer;
- some NCAs may foster interpersonal competition. That is, some or all of the time, they are pursued as a sport (as defined by Coakley, 2001, p. 20). The attraction of the central activity done in nature remains important here, but competing against other people becomes another main reason for engaging in it.

To be as clear as possible about the nature of NCAs, we also need to inventory the activities excluded from the list of NCAs. They are

- activities in which participants are not, themselves, in the focal natural setting. Such is the situation for those who fly kites or model airplanes or run model boats or cars. It is the kite, plane, boat or car that experiences, as it were, the natural element of air, water or land, not the person controlling it from a distance;
- high-risk or 'extreme' activities (defined as such by the participant), primarily because they shift the participant's attention away from the inherent appeal of the core activity in its natural setting to avoiding the substantial possibility of serious injury or death as well as to other extraneous considerations (discussed later in this chapter);
- activities that offer challenges through one of the six elements, but that are undertaken in substantially artificial circumstances. They include those requiring use of swimming pools, ice rinks (indoor and outdoor), artificial climbing walls, fish and game farms, snowboard parks, velodromes, luge and bobsleigh runs, ski jumps and groomed alpine ski and snowboard runs, race tracks (e.g., auto, horse, motorcycle), courses (e.g., track and field, sled dog, model trains and cars), rodeo arenas, pastures (for raising horses, sheep, etc.), golf courses and equestrian grounds;[1]
- a great variety of casual leisure activities that, in some ways, resemble NCAs, but nonetheless fail to meet the criteria of serious leisure, including the criterion of a natural challenge. These include bungee jumping, zip rides, commercial whitewater rafting (experts at the helm excluded), inner tubing on a river, horseback riding (for greenhorns), sledding and tobogganing (on natural terrain), casual swimming and snorkelling;
- activities whose chief appeal is promoting fitness, among them, jogging, swimming, rowing and skating, all done in natural settings. True, these activities may be more attractive when done outside rather than inside, but even then, the principal goal is other than finding fulfilment in the activity itself or enjoying the natural environment in which it occurs;
- the activity of gathering a resource found in nature to be used in a making and tinkering hobby (explained further in Chapter 5).

In this book we examine a wide variety of NCAs. Meanwhile, to sharpen still further initial understanding of this idea, consider the following sample of NCAs:

- Collecting natural objects (shells, leaves, rocks, etc)
- Orienteering

- Back-country skiing and snowboarding
- Caving
- Mushrooming (gathering edibles or scientific specimens)
- Bird watching
- Amateur ornithology, astronomy, botany, entomology, meteorology – the natural challenge is to find new phenomena, rare but known phenomena, predict weather and so on[2]
- Search and rescue, maintaining hiking and cross-country ski trails, volunteering through an NCA
- Sailing
- Wilderness camping.

Bear in mind that we are, in all this, trying to generalize about NCAs and their enthusiasts. Generalizations gloss individual differences, however, such that a small minority of individuals may interpret their own pursuits and experiences in them differently from any given generalization. Thus, some people love fitness activities that most others find disagreeably obligatory, have no sense of being forced to do them and love them even more when done in a pleasant setting out of doors.

What is special about the NCA?

Have we split hairs in our attempt to distinguish NCAs from other outdoor leisure activities? After all, one of the most popular and scientifically researched groups in this vast realm is the one classified as high-risk, or extreme, even though the number of participants here is, by comparison, very small. Far more popular than the high-risk activities are the multitude of casual leisure interests undertaken outside. Finally there are many enthusiasts who go in for outdoor fitness which they regard not as leisure but as disagreeable obligation as well as people who pursue serious leisure activities out of doors in substantially artificial circumstances (see, for example, earlier discussion on alpine skiing and snowboarding).

These neighbouring categories of activity aside, the NCAs stand out as separate and important. First, it appears that many people also pursue NCAs considered as a group, even though we lack precise comparative data with which to substantiate this claim (we return to this question in all remaining chapters). Second, these activities, unlike the others, are exclusively serious leisure, experienced in this instance in highly appealing natural settings. Its principal reward is personally fulfilling execution of a core activity done in an environment that is itself awe-inspiring. Third, many NCAs, when pursued regularly, also contribute to their participants' physical fitness, suggesting that, for them, achieving fitness in such settings may be more appealing than doing so in artificial circumstances.

Fourth, the NCAs offer greater scope for human agency than many other kinds of leisure activity. Thus, swimmers in lakes and rivers with no officially established swimming areas have the freedom to swim where they wish compared with their counterparts who are constrained by official boundaries demarcating them or limited by the edges of artificial pools. Hikers and cross-country skiers may explore off-trail, hang-gliders are free to search wind currents and chart their course of flight at will and hobbyist sailors have an entire ocean or lake to explore (necessarily mindful of dangerous areas and changing weather patterns). Artificial areas constrain such tendencies, while scope for human agency, say participants in many of the activities considered in this book, is a central attraction of their chosen NCAs.

Fifth, unlike artificial conditions, natural settings are constantly changing. Nature, in a sense, becomes a 'player' that participants must try to 'read', predict and adapt to in order to successfully meet a challenge; that is, to varying extents the activity takes place 'on nature's terms'. This constant variability provides an endless fascination for many NCA enthusiasts and the intimate knowledge required of their respective natural elements helps to foster a relatedness and respect for these environments. The deep sense of connection and intimacy that many NCA participants feel in a particular natural setting may help to sustain long-term commitment to the activity, as Davidson (2006, 2008) has shown with mountaineers, who even as they aged and could no longer undertake technical climbs would hike up valleys just to spend time in the mountains.

Nature challenge activity as a concept is a newcomer to the literature on leisure studies. Stebbins (2005a) coined and defined the term to capture how his sample of kayakers, snowboarders and mountain climbers viewed the core activities of their hobbies. It was also clear from this study that there are many more NCAs than these three and, following further analysis of the data from the 2005 study, that the three types of hobbyists also pursue such activities in their natural settings because of the wonder with which they regard those settings.

Awe and wonder

This book explores systematically the proposition that an NCA is pursued in the awe-inspiring natural environment of that activity. Being in awe, or wonder, of an aesthetically attractive part of nature brings a psychological dimension to our study of NCAs. For the kayakers their wonder-filled environment had, among other features, the sound, sight and feel of the rushing mountain rivers and creeks and the rock, earth, trees and vegetation through which they flow. The snowboarders were enthralled with the snow-filled back-country setting through which they rode as it descended, often precipitously, before them. The mountain climbers loved rock, its nooks, crannies and solidity (when present) and the way these qualities and others combined to create a sense of being suspended in air far above the base of the steep slopes they mounted. Borrie and Roggenbuck (2001) reviewed a number of studies that have measured emotion, mood, attention states and feelings of connection with nature. The two authors learned from their research that intense awe of the environment can be variable, depending on what participants are doing, how long they have been doing it, meteorological conditions and the like.

Although we have chosen to summarily describe such feelings with the terms 'awe' and 'wonder', we pondered several alternatives. Two of these were 'euphoria' and 'exhilaration', which though rejected, we nevertheless regard as closely synonymous with awe. Another was 'spirituality', refused by us because of its oftentimes close association with religion. Some people in an NCA might well describe the experience precisely this way, but others would not. 'Delight' would have served our purposes were it not for its sense of pleasure. Casual leisure generates pleasure, whereas the central rewards of serious leisure are deeper. In brief we were trying to communicate with both noun and adjective (awe-inspiring, wonder-filled) the 'wow' feeling – the emotion – of an aesthetic encounter with nature, as experienced during normal conduct of the hobby's core activity.[3]

The theoretic framework guiding the present study – the serious leisure perspective – roots in the concepts of leisure activity, core activity and the experience of pursuing both.

The serious leisure perspective

We start with a definition of leisure. It is an uncoerced, positive activity that, using their abilities and resources, people both want to do and can do at either a personally satisfying or a deeper fulfilling level (Stebbins, 2005b, 2009d). 'Free time' is time away from unpleasant obligation, with pleasant obligation being understood by participants as essentially leisure – *homo otiosus,* leisure man, feels no significant coercion to enact the activity in question. What people experience as leisure revolves around one or more *core activities,* or distinctive sets of interrelated actions or steps that must be followed to achieve an outcome or product attractive to the participants.

The serious leisure perspective may be described, in simplest terms, as the theoretic framework that synthesizes three main forms of leisure showing, at once, their distinctive features, similarities and interrelationships.[4] Additionally the Perspective (wherever Perspective appears as shorthand for serious leisure perspective, to avoid confusion, the first letter will be capitalized) considers how the three forms – serious leisure, casual leisure and project-based leisure – are shaped by various psychological, social, cultural and historical conditions. Each form serves as a conceptual umbrella for a range of types of related activities. That the Perspective takes its name from the first of these should, in no way, suggest that we regard it, in some abstract sense, as the most important or superior of the three. Rather the Perspective is so titled, simply because it got its start in the study of serious leisure; such leisure is, strictly from the standpoint of intellectual invention, the godfather of the other two. Furthermore serious leisure has become the benchmark from which analyses of casual and project-based leisure have often been undertaken. So naming the Perspective after the first facilitates intellectual recognition; it keeps the idea in familiar territory for all concerned.

Serious leisure

Serious leisure is constituted of three types: amateurism, hobbyist activities and career volunteering. Amateurs are found in art, science, sport and entertainment, where they are inevitably linked, one way or another, with professional counterparts who coalesce, along with the public whom the two groups share, into a three-way system of relations and relationships (the professional-amateur-public, or P-A-P, system). A professional is identified and defined economically: one who is paid for the activity in question. This definition puts into relief the possibility that some amateurs and hobbyists may begin to make some sort of living at the activity. Freed partly or wholly from having to make a living in another field, it becomes possible for these people to devote additional time to their serious leisure and thus, in some instances, excel over their counterparts in leisure who can only pursue the activity after a full day's work elsewhere. This intermediate terrain between unpaid leisure and full-time work in an activity is home to many an enthusiast in art, sport, science and entertainment (for further details on this argument, see Stebbins, 2007, pp. 6–7).

Hobbyists lack the professional alter ego of amateurs, although they sometimes have commercial equivalents and often have small publics who take an interest in what they do. Hobbyists may be grouped according to five categories: collectors, makers and tinkerers, activity participants (in non-competitive, rule-based, pursuits such as fishing and barbershop singing), players of sports and games (in competitive, rule-based activities with no professional counterparts like long-distance running and competitive swimming) and the enthusiasts of the liberal arts hobbies. The rules guiding rule-based pursuits are, for the most part, either subcultural (informal) or regulatory (formal). Thus seasoned hikers in the Canadian Rockies know they should, for example, stay on established trails, pack out all garbage, be prepared for changes in weather and make noise to scare off bears. The liberal arts hobbyists are enamoured of the systematic acquisition of knowledge for its own sake. Many of them accomplish this by reading voraciously in a field of art, sport, cuisine, language, culture, history, science, philosophy, politics or literature (Stebbins, 1994). But some of them go beyond this to expand their knowledge still further through cultural travel.

Many NCAs fall under the heading of non-competitive, rule-based activity participation. True, actual competitions are sometimes held in these hobbies (e.g., fastest time over a particular course), but mostly beating nature while pursuing a leisure goal is thrill enough. But there are also hobbies that offer nature challenges largely, if not entirely, free of the practice of formal competition. Some, most notably fishing and hunting, in essence exploit the natural environment. Others centre more fully on appreciation of the outdoors, among them hiking, backpacking, bird watching and horseback riding (Stebbins, 1998a, p. 59).

Turning next to volunteering, Cnaan et al. (1996) identified four dimensions they found running throughout the several definitions of volunteering they examined. These dimensions are free choice, remuneration, structure and intended beneficiaries. The following definition has been created from these four: volunteering is uncoerced help offered either formally or informally with no or, at most, token pay and done for the benefit of both other people and the volunteer (Smith et al., 2006, pp. 239–240). The volunteer provides altruistically a service or benefit to one or more individuals who are not part of that person's family. This is essentially a volitional definition, which contrasts with the more widely embraced economic definition of volunteer as unpaid work.

Concerning the free choice dimension, the language of (lack of) 'coercion' is preferred, since that of 'free choice' is hedged about with numerous problems (see Stebbins, 2005b). The logical difficulties of including obligation in definitions of volunteering militate against including this condition in the foregoing definition (see Stebbins,

2001c). As for remuneration, volunteers retain their voluntary spirit providing they avoid becoming dependent on any money received from their volunteering. Structurally, volunteers may serve formally in collaboration with legally chartered organizations or informally in situations involving small groups, sets or networks of friends, neighbours, colleagues and the like that have no such legal basis. Finally, it follows from what was said previously about altruism and self-interest in volunteering that both the volunteers and those who they help find benefits in such activity. It should be noted, however, that the field of serious leisure, or career, volunteering, even if it does cover considerable ground, is still narrower than that of volunteering in general, which includes helping as casual leisure.

Volitionally speaking, volunteer activities are motivated, in part, by one of six types of interest: interest in activities involving (1) people, (2) ideas, (3) things, (4) flora, (5) fauna or (6) the natural environment (Stebbins, 2009d). Each type, or combination of types, offers its volunteers an opportunity to pursue, through an altruistic activity, a particular kind of interest. Thus, volunteers interested in working with certain ideas are attracted to idea-based volunteering, while those interested in certain kinds of animals are attracted to faunal volunteering. Interest forms the first dimension of a typology of volunteers and volunteering.

But, since volunteers and volunteering cannot be explained by interest alone, a second dimension is needed. This is supplied by the serious leisure perspective and its three forms. This Perspective, as already noted, sets out the motivational and contextual (sociocultural, historical) foundation of the three. The intersections of these two dimensions produce 18 types of volunteers and volunteering, exemplified in idea-based serious leisure volunteers, material casual leisure volunteering (working with things) and environmental project-based volunteering (see Table 1.1).

Serious leisure is further defined by six distinguishing qualities (Stebbins, 1992, pp. 6–8), qualities found among amateurs, hobbyists and volunteers alike. One is the occasional need to *persevere,* such as in confronting danger (Fine, 1988, p. 181), supporting a team in losing season (Gibson et al., 2002, pp. 405–408), or embarrassment (Floro, 1978, p. 198). Yet, it is clear that positive feelings about the activity come, to some extent, from sticking with it through thick and thin, from conquering adversity. A second quality is, as already indicated, that of finding a leisure *career* in the endeavour, shaped as it is by its own special contingencies, turning points and stages of achievement or involvement. Because of the widespread tendency to see the idea of career as applying only to occupations, note that, in this

Table 1.1   A leisure-based theoretic typology of volunteers and volunteering

Leisure interest	Type of volunteer		
	Serious Leisure (SL)	Casual Leisure (CL)	Project-Based Leisure (PBL)
Popular	SL Popular	CL Popular	PBL Popular
Idea-Based	SL Idea-Based	CL Idea-Based	PBL Idea-Based
Material	SL Material	CL Material	PBL Material
Floral	SL Floral	CL Floral	PBL Floral
Faunal	SL Faunal	CL Faunal	PBL Faunal
Environmental	SL Environmental	CL Environmental	PBL Environmental

definition, the term is much more broadly used, following Goffman's (1961, pp. 127–128) elaboration of the concept of 'moral career'. Broadly conceived of, careers are available in all substantial, complicated roles, including especially those in work, leisure, deviance, politics, religion and interpersonal relationships.

Careers in serious leisure commonly rest on a third quality: significant personal *effort* based on specially acquired *knowledge, training, experience* or *skill,* and, indeed, all four at times. Examples include such characteristics as showmanship, athletic prowess, scientific knowledge and long experience in a role. Fourth, eight *durable benefits,* or broad outcomes, of serious leisure have so far been identified, mostly from research on amateurs. They are self-actualization, self-enrichment, self-expression, regeneration or renewal of self, feelings of accomplishment, enhancement of self-image, social interaction and belongingness, and lasting physical products of the activity (e.g., a painting, scientific paper, piece of furniture). A further benefit – self-gratification, or the combination of superficial enjoyment and deep satisfaction – is also one of the main benefits of casual leisure, to the extent that the enjoyment part dominates.

A fifth quality of serious leisure is the *unique ethos* that grows up around each instance of it, a central component of which is a special social world where participants can pursue their free-time interests. Unruh (1980, p. 277) developed the following definition:

A *social world* must be seen as a unit of social organization which is diffuse and amorphous in character. Generally larger than groups or organizations, social worlds are not necessarily defined by formal boundaries, membership lists, or spatial territory.... A social world must be seen as an internally recognizable constellation of actors, organizations, events, and practices which have coalesced into a perceived sphere of interest and involvement for participants. Characteristically, a social world lacks a powerful centralized authority structure and is delimited by... effective communication and not territory nor formal group membership.

In another paper Unruh added that the typical social world is characterized by voluntary identification, by a freedom to enter into and depart from it (Unruh, 1979). Moreover, because it is so diffuse, ordinary members are only partly involved in the full range of its activities. After all, a social world may be local, regional, multiregional, national or even international.

Third, people in complex societies, including especially those in the West, are often members of several social worlds. Finally, social worlds are held together, to an important degree, by semiformal, or mediated, communication. They are rarely heavily bureaucratized yet, due to their diffuseness, they are rarely characterized by intense face-to-face interaction. Rather, communication is typically mediated by newsletters, posted notices, telephone messages, mass mailings, Internet communications, radio and television announcements, and similar means, with the strong possibility that the Internet could become the most popular of these in the future.

The sixth quality revolves around the preceding five: participants in serious leisure tend to *identify* strongly with their chosen pursuits. In contrast, casual leisure, though hardly humiliating or despicable, is nonetheless too fleeting, mundane and commonplace for most people to find a distinctive identity there. In fact, as this study shows, a serious leisure pursuit can hold greater appeal as an identifier than the person's work role.

Rewards and costs

In addition, research on serious leisure has led to the discovery of a distinctive set of rewards for each activity examined (Stebbins, 2007, pp. 13–15). In these studies the participant's leisure fulfilment has been found to stem from a constellation of particular rewards gained from the activity, be it caving, trapping, ice climbing or wind surfing.[5] Furthermore, the rewards are not only fulfilling in themselves, but also fulfilling as counterweights to the costs encountered in the activity.

Put more precisely, then, the drive to find fulfilment in serious leisure is the drive to experience the rewards of a given leisure activity, such that its costs are seen by the participant as more or less insignificant by comparison. This is at once the meaning of the activity for the participant and his or her motivation for engaging in it. It is this motivational sense of the concept of reward that distinguishes it from the idea of durable benefit set out earlier, an idea that emphasizes outcomes rather than antecedent conditions. Nonetheless, the two ideas constitute two sides of the same social psychological coin.

The rewards of a serious leisure pursuit are the more or less routine values that attract and hold its enthusiasts. Every serious leisure career both frames and is framed by the continuous search for these rewards, a search that takes months, and in many sports, years, before the participant consistently finds deep fulfilment in his or her amateur, hobbyist or volunteer role. The ten rewards presented below emerged in the course of various exploratory studies of amateurs, hobbyists and career volunteers (for a summary of these studies, see Stebbins, 2001a). As the following list shows, the rewards of serious leisure are predominantly personal.

Personal rewards

1. Personal enrichment (cherished experiences)
2. Self-actualization (developing skills, abilities, knowledge)
3. Self-expression (expressing skills, abilities, knowledge already developed)
4. Self-image (known to others as a particular kind of serious leisure participant)
5. Self-gratification (combination of superficial enjoyment and deep satisfaction)
6. Re-creation (regeneration) of oneself through serious leisure after a day's work
7. Financial return (from a serious leisure activity)

Social rewards

8. Social attraction (associating with other serious leisure participants, with clients as a volunteer, participating in the social world of the activity)
9. Group accomplishment (group effort in accomplishing a serious leisure project; senses of helping, being needed, being altruistic)
10. Contribution to the maintenance and development of the group (including senses of helping, being needed and altruistically making the contribution)

In the various studies on amateurs, hobbyists and volunteers, these rewards, depending on the activity, were often given different weightings by the interviewees to reflect their importance relative to each other. Nonetheless, some common ground exists, for the studies on sports, for example, do show that, in terms of their personal importance, most serious leisure participants rank self-enrichment and self-gratification as number one and number two. Moreover, to find either reward, participants must have acquired sufficient levels of relevant skill, knowledge and experience (Stebbins, 1979, 1993). In other words, self-actualization, which was often ranked third in importance, is also highly rewarding in serious leisure.

Serious leisure experiences also have their negative side. In line with this reasoning, we have always asked our respondents to discuss the costs they find in their serious leisure, namely, the dislikes, tensions and disappointments. So far, it has been impossible to develop a general list of costs as has been done for the rewards, since the costs tend to be highly specific to each serious leisure activity. Thus each activity studied to date has been found to have its own constellation of costs, but as the respondents see them, they are invariably and heavily outweighed in importance by the rewards of the activity. For instance, amateur astronomers often dislike the cold weather in which their celestial observations must sometimes be made or the need to drive long distances to find an open field and low light pollution from which to observe. Yet they still tend to regard this activity as highly fulfilling – as (serious) leisure – because it also offers certain powerful rewards, one of them being the awe-inspiring sense of 'getting lost in the heavens'.

Finally, it has been argued over the years that amateurs and sometimes even the activities they pursue are marginal in society, for amateurs are neither dabblers nor professionals (see Stebbins, 1979). Moreover, studies of hobbyists and career volunteers show that they and some of their activities are just as marginal and for many of the same reasons (Stebbins, 1996, 1998b). Several properties of serious leisure give substance to these observations. One, although seemingly illogical according to common sense, is that serious leisure is characterized empirically by an important degree of positive commitment to a pursuit (Stebbins, 1992, pp. 51–52). This commitment is measured, among other ways, by the sizeable investments of time and energy in the leisure made by its devotees and participants. Two, serious leisure is pursued with noticeable intentness, with such passion that Erving Goffman

(1963, pp. 144–145) once qualified amateurs and hobbyists as the 'quietly disaffiliated'. People with such orientations towards their leisure are marginal compared with people who go in for the ever-popular forms of casual leisure.

Psychological flow

Although the idea of flow originated in the work of Mihalyi Csikszentmihalyi (1990) and has, therefore, an intellectual history quite separate from that of the Perspective, it does nevertheless happen on occasion that it is a key motivational force in serious leisure. Indeed flow has now been conceptually integrated in the Perspective, where it plays a distinctive role (Stebbins, 2007, pp. 15–17; Stebbins, 2010). Concerning this book, flow is highly prized in a number of the NCAs we will examine. What then is flow?

Flow, a form of optimal experience, is possibly the most widely discussed and studied generic intrinsic reward in the psychology of work and leisure. Although many types of work and leisure generate little or no flow for their participants, those that do are found primarily in the 'devotee occupations' (Stebbins, 2004a) and in serious leisure. Still, it will be evident that each work and leisure activity capable of producing flow does so in terms unique to it. And it follows that each of these activities must be carefully studied to discover the properties contributing to its distinctive flow experience.

In his theory of optimal experience, Csikszentmihalyi (1990, pp. 3–5, 54) describes and explains the psychological foundation of the many flow activities in work and leisure, as exemplified in chess, dancing, surgery and rock climbing. Flow is 'autotelic' experience, or the sensation that comes with the actual enacting of intrinsically rewarding activity. Over the years, Csikszentmihalyi (1990, pp. 49–67) has identified and explored eight components of this experience. It is easy to see how this quality of work, when present, is sufficiently rewarding and, it follows, highly valued to endow it with many of the qualities of serious leisure, thereby rendering the two inseparable in several ways. And this even though most people tend to think of work and leisure as vastly different. The eight components are

1. sense of competence in executing the activity;
2. requirement of concentration;
3. clarity of goals of the activity;

4. immediate feedback from the activity;
5. sense of deep, focused involvement in the activity;
6. sense of control in completing the activity;
7. loss of self-consciousness during the activity;
8. sense of time is truncated during the activity.

These components are self-evident, except possibly for the first and the sixth. With reference to the first, flow fails to develop when the activity is either too easy or too difficult; to experience flow the participant must feel capable of performing a moderately challenging activity. The sixth component refers to the perceived degree of control the participant has over execution of the activity. This is not a matter of personal competence; rather it is one of degree of influence of uncontrollable external forces, a condition well illustrated in situations faced by some of the hobbyists in this study, such as when the water level suddenly rises on the river or an unpredicted snowstorm results in a whiteout on a mountain snowboard slope.

Flow is a cardinal motivator in such NCAs as mountaineering, scuba diving and hang gliding, even if it is only an occasional state of mind there. That is, in any given outing in one of these hobbies, participants only experience flow some of the time. Meanwhile it is not even this central in some other outdoor hobbies. It is only sporadic in, for instance, mountain scrambling, backpacking and horseback riding. By contrast, it is certainly a motivational feature in, say, mountain biking as well as cross-country and back-country skiing. It should be noted, however, that, by its very nature, flow tends to dilute somewhat the sense of environmental wonder participants feel, because their concentration is so intensely focused on executing the core activity.

Casual leisure

Casual leisure, which in comparison with serious leisure is considerably less substantial and offers no career, was defined in the previous section as an immediately, intrinsically rewarding, relatively short-lived pleasurable core activity, requiring little or no special training to enjoy it (Stebbins, 1992, 1997, p. 18). Over the years eight types have been identified. They include:

- play (including dabbling),
- relaxation (e.g., sitting, napping, strolling),

- passive entertainment (e.g., TV, books, recorded music),
- active entertainment (e.g., games of chance, party games),
- sociable conversation (e.g., gossip, 'idle chatter', banter),
- sensory stimulation (e.g., sex, eating, drinking, sight seeing),
- casual volunteering (e.g., handing out leaflets, stuffing envelops), and
- pleasurable aerobic activity.

The first six types are more fully discussed in Stebbins (1997), while casual volunteering is considered further in Stebbins (2003). The last and newest addition to this typology – pleasurable aerobic activity – refers to physical activities that require effort sufficient to cause marked increase in respiration and heart rate. Here 'aerobic activity' is treated of in the broad sense, as all activity calling for such effort, which to be sure, includes the routines pursued collectively in (narrowly conceived of) aerobics classes and those pursued individually by way of televised or video-taped programmes of aerobics (Stebbins, 2004c). Yet, as with its passive and active cousins in entertainment, pleasurable aerobic activity is, at bottom, casual leisure. That is, to do such activity requires little more than minimal skill, knowledge or experience. Examples include the game of the Hash House Harriers (a type of treasure hunt held in the outdoors), kickball (described in *The Economist,* 2005, as a cross between soccer and baseball), and such children's games as tag and hide-and-seek.[6]

This brief review of the types of casual leisure reveals that they share at least one central property: all are hedonic. More precisely, all produce a significant level of pure pleasure, or enjoyment, for those participating in them. In broad, colloquial language, casual leisure could serve as the scientific term for the practice of doing what comes naturally. Yet, paradoxically, this leisure is by no means wholly frivolous, for there are some clear benefits in pursuing it. Moreover, unlike the evanescent hedonic property of casual leisure itself, its benefits are enduring, a property that makes them worthy of extended analysis in their own right (discussed further in Stebbins, 2007, pp. 41–43).

In this book on NCAs, casual leisure demarcates one of their boundaries. True, pleasurable aerobic activity, for example, occurs in natural settings which may be awe-inspiring. The same may be said for sightseeing. Yet neither is serious leisure, since effort and challenge are minimal. It is likewise for dabbling in parachuting, water skiing or fishing; people sometimes try these activities once or twice simply for the experience or to say they have done them. This is done without intending to 'get good at them'.

Project-based leisure

Project-based leisure is a short-term, reasonably complicated, one-off or occasional, though infrequent, creative undertaking carried out in free time, or time free of disagreeable obligation (Stebbins, 2005c). It is for all that neither serious leisure nor intended to develop into such. Examples include surprise birthday parties, elaborate preparations for a major holiday and volunteering for sports events. Though only a rudimentary social world springs up around the project, it does in its own particular way bring together friends, neighbours or relatives (e.g., through a genealogical project or Christmas celebrations), or draw the individual participant into an organizational milieu (e.g., through volunteering for a sports event or major convention).

A main difference separating project-based leisure from serious leisure is that the first, being short-term, fails to generate a sense of career. Otherwise, however, there is here need to persevere, some skill or knowledge may be required and, invariably, effort is called for. Also present are recognizable benefits, a special identity, and often a social world of sorts, though it appears, one usually less complicated than those surrounding many serious leisure activities. And perhaps it happens at times that, even if not intended at the moment as participation in a type of serious leisure, the skilled, artistic or intellectual aspects of the project prove so attractive that the participant decides, after the fact, to make a leisure career of their pursuit as a hobby or an amateur activity. Project-based leisure is also capable of generating many of the rewards experienced in serious leisure.

Project-based leisure sets another boundary of the NCA as a type of leisure. It is certainly possible to undertake projects in wonder-filled natural settings, including a one-time afternoon of guided ocean fishing or hot-air ballooning or a guided wilderness camping trip. These projects would have, as such, been anticipated in advance by booking ahead the appropriate service and putting aside the money and time needed to use it. A participant might also read about ocean fishing or hot-air ballooning prior to the occasion. In the end, however, such projects are not serious leisure.

Risk and adventure

Having now set out the theoretic framework for the present study, it is in order that we consider, using the Perspective as background, two ideas that fail to appear there: risk and adventure. There is a substantial scientific (not to mention popular) literature on both, and it is reasonable to ask why they have been excluded from the serious leisure perspective and from this book on NCAs.[7]

Being in flow in these activities, as opposed to experiencing leisure there more generally, presupposes *manageable* challenge, nothing too easy, which is boring, but nothing too difficult, which in some NCAs can be terrifying (the second meaning of awe). Flow is most intense when people operate at or near their mental and physical limits, but stop short of going beyond them. Additionally, whether someone is in flow depends on that person's background of experience, level of native talent as well as acquired skill and knowledge in the activity. High loadings on these criteria raise the mental and physical limits; that is they boost the level of manageability. At the same time this gives some non-participants who are watching, or reading about, the participant in action the impression that the latter is facing danger and high risk. But the participant is not of this opinion.

This is, in essence, why we reject the idea of high physical risk – as participants define this for themselves – as a general descriptor of NCAs. For them high physical risk in their NCAs has to do with reasonable likelihood of perceived danger at their level of experience, talent and so on that holds out significant probability of major injury, even death. Our NCA enthusiasts undertake their serious leisure to beat nature in a particular way, some of them pushing their limits to do this, and from time to time to experience flow as a result. Taking great risk, as they define it, abruptly and dramatically diminishes the feeling of flow, and more generally, that of leisure, for such risk emerges unexpectedly, feeding on the sense that the activity has now become frighteningly unmanageable. With this change in meaning of the activity goes one cardinal reason for doing it in the first place. In vernacular terms, it is no longer fun; now the psychological dimension of their NCA has lost its appeal.

Now, it is true, as Donnelly (2004, p. 44) has observed, that people may, in response to peer pressure, take unacceptable risks (as personally defined) in sports of the kind being considered here. They may also suddenly find themselves in greatly risky situations not of their making, as caused by unexpected changes in weather or conditions of the course (e.g., sudden down draft while hang gliding, exceptionally strong wind while sailing, a 'white out' of snow while skiing downhill) or unforeseen failure of equipment. Faced with such contingencies the individual may be unable to leave, transforming into coerced activity what was previously considered (un-coerced) leisure. This is not, however, why they are so strongly drawn to the activity. Indeed, such experiences could lead some of them to abandon it in the future, as happened to a few of the participants in the mountain hobbies study (Stebbins, 2005a).

Concerning high risk in the area of nature challenge activity, there are, when seen through the eyes of participants, at least four types (Stebbins, 2005a, pp. 18–21). *Unmanaged high risk,* or risk that emerges only when people lose concentration, get fatigued or otherwise suddenly become unable to draw on their acquired skills, knowledge and experience that keep them in leisure and, at its peak and where possible, in flow. The second is *fortuitous high risk,* or greatly improbable risk from uncontrollable sources such as snow avalanches, falling rock, sudden elevations of water level (as caused by heavy rain upstream) and ice slicks (on cross-country ski trails). Mountaineers call these 'objective hazards', hazards created by nature. Consider the following example:

> Two Canmore sisters [not mountaineers] were collecting rocks on a family hike in the Rookies when a huge gust of wind toppled an 18-metre spruce tree on top of them, their father told the *Herald* on Tuesday.
>
> The accident happened at about 2:30 p.m. Monday at Marble Canyon, a hiking trail on Highway 93S, about 15 kilometres southwest of Castle Mountain in Kootenay National Park.
>
> Parks staff said winds in the area were extremely high.
>
> Several trees were reported to have blown over on another path in the area and in a nearby campground, said Shelley Humphries, spokeswoman for Parks Canada.
>
> 'It not unusual for trees to area to fall (in the area),' said Humphries. 'What is unusual is that some one was hurt'.
>
> With their parents, Peggy and Paul, the sisters were walking across the upper bridges of the canyon when the tree fell on them.
>
> (Poole, 2003, p. A1)

Social high risk, the third type, was presented earlier as pressure from peers to engage in the activity at a level regarded by the pressured participant as risky; that person sees it as going significantly beyond his or her acquired skills, knowledge and experience. But social risk can also include intentionally taking great risks for the fame and perhaps even the fortune that it brings. As example of this latter orientation consider that of Régine Cavagnoud, French world champion in alpine skiing, who died in a collision with a ski coach while hurtling down a slope in the Alps.

> Many times previously Miss Cavagnoud had been badly injured on the slopes while pushing herself to her natural limits, and probably beyond, in her drive to become a world champion.... Miss Cavagnoud did feel fear. Considering the risks involved, there have been relatively few deaths on the slopes.... But many skiers are badly injured. Miss Cavagnoud dreaded ending up in a wheelchair. But even more, she said, she dreaded doing badly.
>
> (*The Economist,* 2001)

A number of popular books also celebrate taking intentional social risk in nature, thereby contributing disproportionately to the commonsense notion that the hobbies in question are inherently hazardous (e.g., Sebastian Junger, *The Perfect Storm* [1999]; Jonathan Shay, *Achilles in Viet Nam* [1995]). At the same time many NCA participants can be highly critical of this type of risk taking and its potentially disastrous consequences (e.g., Jon Krakauer, *Into Thin Air* [1997]; see also Davidson, 2006).

Additionally high risk of the social variety may be prized for its capacity to generate an 'adrenaline rush'. One survivor of a snowmobile high marking event (see Chapter 6) that was terminated by a massive avalanche said when asked why he takes such risks: 'Adrenaline.... The feeling is so good. It's like a drug.... It's all about adrenaline' (Staseson & Komarnicki, 2010, p. A5). Two men were killed in the avalanche with around 30 more being injured, some seriously.

And sometimes nature challenge hobbyists take social risks because others in their group want to press on (e.g., up the mountain face, down the river rapids). This they do in circumstances where, though their activities are subject to the first two types of risk, more experienced members of the group, with careful preparation and advance information, usually know how to avoid them and, in fact, usually do. They are avoided because, in significant part, even moderate risk seriously dilutes the senses of flow and leisure. It is questionable whether either is felt at all in social risk, though that may matter little in any case, since the object here is to establish an identity as 'gutsy', as a devil-may-care individual, or whose motive is fame and fortune, if not both of these. That is some enthusiasts who go in for high-risk activity are literally paid to engage in it by a sponsor looking for a sensational promotion of a product, possibly augmented by further remuneration through public speaking engagements or a contracted book or article on the subject. The most celebrated might even get recognized in *Guinness World Records*. For an example of group pressure leading to social high risk, we hear from a female kayaker about her early experiences in her hobby:

> R: I did more than I should have done. Spent a lot of time on the Kananaskis [River], and then paddled sort of on even some class-3 stuff. But a lot of the time I was very discouraged kayaking. I was simply being dragged along by these guys. I'd go on their kayaking trips, and we would go to the Clearwater [River], which scared the hell out me, but everybody was doing it.
>
> I: Did you just get out and portage on that?
>
> R: No I swam a lot [bailed out of her kayak]. I portaged a lot, too, but mostly I just swam. So the first summer was a lot on the Kan[anaskis] and a few trips elsewhere to bigger rivers, but not that I was necessarily confident on most of them. I was o.k. But I couldn't roll [the kayak], so that was a big thing.
>
> (taken from Stebbins, 2005a, p. 20)

Finally, included in this list is what Stebbins has labelled *humanitarian high risk*. For example, Jennifer Lois (2003) studied 'peak volunteers' – viewed in the present book through the prism of the serious leisure perspective as leisure participants – who sometimes must challenge nature while carrying out search and rescue missions. Desmond (2007) has studied wildland fire fighters from a similar angle. Peak volunteers seek high risk not for its own sake, but as something they must confront (presumably reluctantly) in their mission to save lives or recover bodies. Such volunteering illustrates well NCA in the area of career volunteering (see Chapter 7).

In short, while the public, in general, and the press, in particular, are fond of describing some of the NCAs in dramatic, global terms as high-risk, or 'extreme', pursuits, this is not, according to our data, how the vast majority of participants in those pursuits see their routine involvement there.

Adventure

The idea of 'adventure' suffers at least as much from imprecision as does that of 'high-risk'. The first means many things to many different people, is undergoing a sort of commercialization at the hands of the tourism and media industries and generally fails, in any case, to describe the essence of so-called adventurous leisure activities, including those done in nature. This said attempts at a scientific definition of adventure do exist.

For instance Johnson (2003) holds that

> the term 'adventure' conjures up images of testing oneself against a challenging environment. This might include exotic and/or potentially dangerous sites such as mountains, rivers and wilderness areas. Although danger is not a necessary component, the term does imply the non-ordinary. Adventure also focuses on the specific activity that the person engages in.

He goes on to observe that an adventure need be neither dangerous nor occur in a challenging physical environment. Whatever else it is adventure is also a state of mind that accompanies involvement in certain challenging activities.[8]

From this statement it is evident that the concept of adventure, scientifically defined, overlaps those of high-risk and NCA. That is not all adventures are risky and some adventures spring from meeting challenges that are artificial such as indoor climbing walls, velodromes and paper targets for marksmanship. For example Steve Fossett, identified by the *Economist* (2008) as an adventurer and record-breaker, was, by his own assessment, not a risk taker. Among his many exploits was participation in the adventurous but low-risk activities of running in the Boston Marathon, swimming across the English Channel, and riding a dogsled in the Iditarod race in Alaska. Nevertheless the idea should be considered part of our emerging grounded theory of NCAs, particularly in its propositions about challenge, experience, activity and state of mind. A sense of adventure can enrich the psychological dimension of an NCA. The concept sensitizes us to the degree of challenge, in that the feeling of adventure grows with the level of natural test confronting the participant.

A study of NCA

Since the concept of NCA has a very recent origin, scientific literature on the idea is, in general, exceedingly slim. Hence the study reported here is, of necessity, exploratory (Stebbins, 2001b). It is designed to generate grounded theory (Glaser, 1978; Glaser & Strauss, 1967) on as wide range of NCAs as we could identify and find information on. The data are exclusively of the 'library' variety: where available we examined past research on particular NCAs as well as drew on biographic and autobiographic materials reporting personal involvements in them. In addition to the limited scholarly sources of journal articles, book chapters and research monographs, we looked primarily in books, videos, newspapers, magazines, newsletters and in such online sources as websites and Web 2.0 (Facebook, My Space, YouTube, etc.).

Our goal was to read about the basic features of and personal experiences in each NCA, whether old or new, in the six elements, always with an eye to learning about the distinctive properties of particular nature challenges as well as their distinctive awe-inspiring qualities. In the main our analysis proceeded along lines of the six previously mentioned dimensions (environmental, geographic, economic, aesthetic, social and psychological). Additionally, and where possible, we always tried to extract from these readings how the NCA being considered met the distinguishing qualities of serious leisure. It is imperative in studying serious leisure to determine that the core activities being examined are neither casual nor project-based leisure. In the end we were able to gather workable amounts (enough for generating valid inductive generalizations) of scholarly and popular material on several hundred NCAs and types. Nevertheless, in a pioneering effort of this sort, it is certainly possible that we have overlooked some new NCAs, and possibly even a few that are more established. Space limitations also forced us to leave out a few activities. Thus the best we can claim is that our sample is representative, rather than exhaustive.

Each activity will be described and examined as thoroughly as possible along the aforementioned lines. This is accomplished as follows: activities undertaken in air are covered in Chapter 2, those pursued in water in Chapter 3, those pursued on land in Chapter 4, those involving flora or fauna in Chapter 5, and those involving ice or snow in Chapter 6. Chapter 7 concentrates on mixed activities, or those substantially based on two or more of the six elements (e.g., multisport and adventure racing) and volunteer NCAs which occur across the spectrum of natural settings. In Chapter 8, our conclusion, we pull together the many generalizations that have emerged in the earlier chapters into as much of a grounded theory about NCAs as our data allow. We accomplish this by addressing ourselves to four sets of implications issuing from our findings: those bearing on theory, policy, consumption and the environment. The intention throughout is to lay the groundwork for further systematic exploration of this new field of study.

Conclusion

Challenging nature is a main way in which contemporary *homo otiosus* takes his leisure in Western society. As indicated people in these societies have differential access to these activities and, even where access is assured, they have different tastes and talents for those activities. Moreover, some people, many of them city dwellers, have no interest at all in any NCA. But, to understand better those who do, we have offered in this chapter a theoretic framework – the serious leisure perspective and the allied concepts of nature challenge, NCA, awe/wonder, high-risk and adventure – with which to guide open-ended study in this scientifically neglected area of free-time activity. The process of examination is extended further by orienting inquiry towards, among other relevant areas, the six analytic dimensions. We then address ourselves to four sets of implications issuing from our findings: those bearing on theory, on policy, on consumption and on the environment.

To the extent that we have been able to locate usable data, we will look at both old and new NCAs. One of the remarkable characteristics of this area at present is the alarming rate and variety of new activities being invented and pursued (Stebbins, 2009a). Also noteworthy is how quickly those that catch on are publicized and globalized, primarily by way of the mass media and the Internet. This is no more evident than in the activities whose principal challenge is gliding or falling through the earth's atmosphere.

2
Air

The most ethereal of the natural elements, air has had an enduring fascination for *homo otiosus*. While we have long dreamt of becoming airborne, the scientific and technological advances that would finally make this a reality are a mere few hundred years old. But thanks to these – and a number of very recent innovations – the natural challenges now available to us in the air range from the exhilarating plunge of a skydive to the serenity of a balloon flight, from the bird-like soaring of a hang glider to the playful manoeuvrability of an autogyro. These activities have opened up a whole new natural playscape of air currents, thermals, storm fronts and mountain waves which we can now share with the birds. Indeed aeronauts often express a kinship with birds, with their freedom and mastery of the skies, and encounters of the feathered kind are a feature in many of their stories.

Despite the sense of freedom often associated with flight, this is the most highly regulated group of NCAs, with most participants requiring licences and being subject to air laws and other bureaucratic processes. Recreational flight also requires a relatively high outlay in terms of equipment and training to acquire the technical expertise necessary for even novice participation. These factors have limited participant numbers to an extent, but more recent trends have been towards forms of flight for which cost and regulations can be kept to a minimum.

For the majority of NCAs included in this chapter the participants are immersed in the air; that is, they leave the ground and stay aloft for sustained periods of time with the assistance of some form of aircraft which enables them to engage with the dynamic forces or gaseous qualities of the air. For these people, the allure of the air is to float, soar, hover, glide and/or fly. However, we have also included two activities for which the sky is the setting and the qualities and characteristics of the air, or more precisely, of various atmospheric and celestial phenomenon, provide the natural challenge, but whose devotees can remain with two feet firmly on the ground while their eyes are turned sky-ward. These are the amateur astronomers and meteorologists.

There are other NCAs, of course, such as kitesurfing, snow kiting and sand yachting, in which participants exploit the qualities of air as a means of propulsion and sometimes even temporary lift. In these cases we have categorized them according to the element over which they travel, but they are clearly 'hybrid' NCAs in which more than one element plays a part in the challenge.

Aircraft can be categorized as either aerostats (deriving lift from lighter-than-air gases) or aerodynes (deriving lift from aerodynamic forces), powered or unpowered. They may have fixed wings, flexible wings or rotary wings. The World Air Sports Federation/Fédération Aéronautique Internationale (FAI) is the global body for promoting aeronautical and astronautical activities. Established in 1905 and based in Lausanne, Switzerland (formally in Paris), it has around 100 member countries and its various sport-specific commissions oversee competition rules, recognize world records and address issues of relevance and concern to members.

We begin with the balloon as it was the first technological breakthrough that achieved free flight. In doing so, the balloon provided crucial evidence of human capabilities and potentialities in the atmosphere, as well as the impetus and opportunity to develop subsequent aerodynamic technologies. The first of these was the parachute, which has led to the modern day NCAs of sky diving, base jumping and wing-suit flying. Our discussion then turns to the unpowered aerodynes – paragliders, hang gliders and gliders/sailplanes. The first section on powered flight covers microlights which are a bridge between the 'free flight' disciplines of paragliding and hang gliding and conventional light aircraft, in terms of design, flying technique and their particular attraction as NCAs. Next are the aeroplanes (or light aircraft) and the rotorcraft, including the increasingly popular rotary-winged microlight, the autogyro. We finish on the ground with the amateur astronomers and meteorologists and their take on the awe-inspiring phenomena within and beyond the earth's atmosphere.

Ballooning

Ballooning marks the beginning of manned, free flight in human history. The vital scientific discoveries that facilitated this were firstly that air was 'stuff', and then that it was made up of different gases of varying densities (Acton, 2002). While their understanding of the science involved was far from perfect, the Montgolfier brothers in France were the first to exploit these characteristics in a manner that could take aeronauts into the air. The first flight of the Montgolfier hot-air balloon in September 1783 had a sheep, duck and rooster on board due to initial concerns about the effects of altitude and a fear of suffocation (Acton, 2002). When the animals survived relatively unscathed, two young French aristocrats – Pilatre de Rozier and Marquis d'Arlandes – volunteered for the first manned flight later that year.

Ten days after the second Montgolfier flight, a rival French balloonist successfully launched a hydrogen-filled balloon. Of the two men aboard this flight, Acton (2002, pp. 50–51) writes:

> Anyone who has ever been in a balloon will know something of the feelings of elation that swept through Charles and Robert at that moment. As they drifted on the breeze, surrendered to and therefore at one with it, it seemed as if they were hanging motionless, while the Earth flew and twisted down beneath them.

It is perhaps impossible to overstate the novelty and potential of the view that was opened up to these early pioneers of flight. Pilot William Langewiesche (1998) argues that the greatest gift of flight is that it enables us to see the world from above and thereby facilitates a radically different perspective on ourselves and our landscape. This 'aerial view', as he terms it, 'carries with it the possibility of genuinely free movement, and allows just the right amount of participation with the landscape – neither as distant as an old fashioned vista nor as entrapping as a permanent involvement' (Langewiesche, 1998, p. 26).

Despite the allure of this vista, ballooning did not catch on as a recreational pursuit with widespread appeal until the second half of the twentieth century when certain technological advances made the hot air design more practical. In the 1950s Paul Edward Yost adapted the Montgolfiers' original design for use by the US Navy. He added a small propane burner, new envelope material (the Montgolfiers' had been made of paper and canvas), came up with the inverted tear shape as the most efficient, added safety features and a new inflation system. In the early 1960s he began producing balloons for recreational use (Ludwig, 1996).

Balloons can fly when filled with any gas that is less dense than air, thus providing the necessary lift. Options include heated air, as in the

Montgolfier design, hydrogen, helium, cooking gas and ammonia gas (Ludwig, 1996). For recreational use the hot-air balloon is the most popular as it is less expensive to purchase and to run than a gas balloon, and easier to rig. The envelope is made from sewn panels of nylon or polyester cloth, which is lightweight but sufficiently strong, and typically there is a vent at the top of the envelope which allows for the release of hot air when required to slow an ascent or facilitate a descent, and for deflation on landing. Baskets are mostly made from woven wicker or rattan and carry the fuel tanks, which are connected by hoses to the burner attached to a frame above the pilot's head. Small, one-person balloons without baskets are known as Hoppers or Cloudhoppers. Most balloons can carry three to five people, while commercial operations with sufficiently large envelopes and baskets can carry upwards of 15 people. Basic instrumentation usually includes an altimeter, a vertical speed indicator (variometer), gauges to indicate fuel quantity and pressure, and sometimes an envelope temperature gauge. An aircraft radio can be used to communicate with the chase crew and, if necessary, with air traffic control and other aircraft.

Balloons are static within the air mass that surrounds them, hence the term *aerostat*. The only means of accelerating, decelerating or influencing the direction of a balloon is to adjust its altitude – by controlling the heat of the air in the envelope – thereby taking advantage of varying air currents. Indeed, it is the ability to successfully manoeuvre within and exploit the movement of air that provides the core natural challenge for balloonists. This requires a thorough understanding of and ability to read wind conditions, including knowledge of the ways in which physical features of the landscape affect the speed and direction of air currents. Navigation is another requirement, and while basic map reading skills will suffice in fine weather, some pilots like to also carry a GPS. The distance covered depends on wind speed and standard balloons may travel up to 40 km in one flight.

Ballooning is by nature a team sport as it generally takes a number of people to rig and launch a balloon and while in flight a chase crew will follow on the ground to retrieve it once it has landed. The best flight conditions are provided by high pressure systems, with light winds on the ground and moderate winds higher up (Ludwig, 1996). For these reasons, the most popular time to fly is within three hours of sunrise or sunset when surface air is at its calmest and winds generally are more predictable. Similarly, light winds make the best landing conditions as balloons have no means of braking their forward motion (Ludwig, 1996). The grace and dignity of the balloon in flight is quickly lost when it comes into land, frequently bumping and dragging its occupants along the ground until it comes to a stop. And while one of the charms of the balloon flight is that one never knows exactly where it will finish, finding a suitable landing place, free of obstacles and easily damaged crops, can be challenging. Overhead power lines pose one of the greatest hazards to balloonists (Smith & Wagner, 1998).

Despite the subsequent innovations of heavier-than-air flight, the lasting fascination with and recreational popularity of the balloon is testament to its peculiar natural challenge and aesthetic appeal. Because a balloon moves with the wind its passengers do not feel or hear a breeze and the experience is described as peaceful and serene. As Carey (1994, p. 123) puts it:

> ... there can be few more pleasant ways of experiencing the joys of flying than in a hot-air balloon in the early morning. The sound of bird-song can usually be heard clearly in the still and quiet dawn air as the huge nylon balloon drifts slowly over mist-laden fields: The roaring flame of the burner is the only intrusion into a peaceful world.

Some countries, for example Britain and the United States, require amateur balloonists to acquire a licence, which generally involves minimum flying experience, practical and written tests covering operation of the balloon, navigation, meteorology, relevant air regulations and human performance and limitations when flying (Carey, 1994). The International Ballooning Commission (CAI), which is part of the FAI, awards badges for outstanding achievement in ballooning, organizes international competitions and ratifies world records, which are held for altitude, duration, distance and shortest time around the world. In competition, the skills of balloonists are tested in tasks which require them to fly to a target upon which they must drop a marker, or 'baggie'. Accuracy depends on the ability to observe micrometeorological conditions and control the balloon in a manner which takes advantage of these (Ludwig, 1996).

Since the 1960s, hot-air ballooning has seen significant growth as an NCA, from a handful to around 10,000 balloons worldwide in the late 1990s (Smith & Wagner, 1998). The number of balloons registered in the United Kingdom has grown from 500 in 1984 to around 1800 in 2010, although the UK's Civil Aviation Authority (CAA) estimates that about half of these are currently active (http://www.caa.co.uk/docs/56/REG%20TOTALS%20WG%20010110.pdf, retrieved 11 April

2010). It is estimated that there are more than 3500 balloons and 4000 licensed pilots in the United States (http://www.hotairballooning.com/faq/entry/How_many_balloons_are_there_9.html, retrieved 16 March 2010).

A hot-air balloon costs about the same as a car or a boat, with popular sport size balloons costing around US$25,000 or more. Ballooning clubs throughout the world arrange events for members and produce publications, such as the British Balloon and Airship Club's bi-monthly journal *Aerostat*. Annual balloon festivals and fiesta are important to the social world of balloonists, as well as providing a spectacle for the interested public.

Parachuting or sky diving

Early umbrella-shaped precursors to the parachute were a source of entertainment in China and in medieval Europe (Crawford, 1996). The first European references to the concept of a parachute appeared in a Leonardo da Vinci sketch in the late fifteenth century, but it was not until 1783, the same year as the first balloon flight, that Louis- Sébastien Lenormand began experimenting with a prototype he called a *parachute* to break the fall of animals dropped from a height (Acton, 2002). Indeed the development of the balloon facilitated and gave impetus to the development of the parachute, the utility of which was seen in its potential to provide a backup in the case of an aborted flight. By 1797 the first human – André-Jacques Garnerin – felt sufficiently confident to jump (voluntarily) from a balloon, using a silk canopy with supporting poles to hold it open. While the attempt was a success, the parachute was very unstable and oscillated wildly from side to side. Stability was improved in the nineteenth century with the addition of a vent in the apex of the canopy allowing the air pressure to be reduced, but for the rest of that century and well into the next, the parachute was primarily used as an amusing stunt for crowds at exhibitions (Acton, 2002). The next major developments in the late nineteenth and early twentieth centuries were the removal of the stiffening poles and the creation of a harness (to replace the baskets used on earlier models), the coatpack that allowed the parachute to be worn on the back and the rip cord (Carey, 1994; Crawford, 1996). These improvements were exploited for military purposes beginning in the First World War, with their effectiveness fully realized by paratroopers during the Second World War. The first parachuting competitions were held in the 1930s, however recreational sky diving truly took hold after the Second World War as the free-time pursuit of ex-military paratroopers (Laurendeau & Sharara, 2008, p. 30).

A sky diver has two chutes – a main and a reserve – packed in a specially designed backpack. Also generally worn are specialized jumpsuits (designed according to the specific sky diving discipline), helmets, goggles and altimeters. Additional equipment may include an automatic activation device, which is capable of calculating altitude and speed of descent, and will activate either of the two chutes at a preset altitude. The rectangular *ram-air* parachute has largely replaced the traditional round canopy as it is more stable and manoeuvrable. The self-inflating ram-air is similar to the paraglider's wing described in the next section, but for the purposes of sky diving it is designed to withstand the stress of being deployed at speeds exceeding 160 km/h, while paragliders have greater lift and range (http://www.nzpf.org/what-is-parachuting-or-skydiving, retrieved 18 March 2010).

Sky divers jump from slow-moving aeroplanes at an average height of 3000 m, but ranging up to 4600 m (Lipscombe, 1999). If the sky diver delays opening the parachute after jumping he or she will accelerate to terminal velocity within seconds, and thereby experience *free-fall* until the parachute is deployed (Crawford, 1996). The skills required of a sky diver include an 'intimate knowledge of the air, and perfect body control' (Carey, 1994). By adopting the full spread or stable position in which the body is arched and the limbs spread in a cross or star shape, divers ensure that they fall face first and do not spin (Carey, 1994). In free-fall, which lasts only about a minute, there is little sensation of falling, instead it is described as a feeling of 'floating' or 'lying on a cushion of air' (Celsi et al., 1993, p. 8). During a jump, a diver's awareness is focused on the 'realities of the jump: body movement; body sensations; body orientation; working with others; and landing' (Lipscombe, 1999, p. 282)

In free-fall divers monitor an altimeter to determine when to open their chutes. Six hundred metres is the minimum for advanced divers. Once the parachute is deployed, its speed and direction can be controlled by toggles attached to steering lines. Ram-air parachutes can glide long distances and at high speeds, and while some jumpers will enjoy the peaceful and relaxing experience of 'flying' their canopies, and some may even perform acrobatic manoeuvres, others view this activity as merely a means of getting safely to the ground (Celsi et al., 1993). With modern parachutes soft, stand-up landings can be achieved with relative ease (Crawford, 1996), the main challenge being to land where planned and to avoid hazardous obstacles such as trees, power lines and bodies of water.

In terms of their knowledge of the air, wind conditions are the most important factor for sky divers. If the wind is blowing, its speed and direction needs to be taken into account so as to return to the designated landing area. 'Spotting' is the skill of determining the location above the ground from which the wind will assist the sky diver to make an accurate landing. Jumpers may check the 'winds aloft' forecast, as well as observing cloud movement, wind socks and flags on the drop zone in order to assist them with this decision.

Lipscombe (1999, p. 268) summarizes the experience of sky divers as follows:

> The exhilaration of the free-fall, the relaxation, the peace and tran-quillity, the silence of the parachute ride, and the physical beauty of the landscapes, give rise to a collection of indescribable feel-ings ... Fear, terror, joy, exhilaration and determination is reported by those who jump to be experienced simultaneously.

These experiences, he continues, produce 'lasting periods of intense emotional uplift referred to by participants as a buzz, a body rush, a high, an adrenaline charge, moments of wonder, of bliss, and some-thing very special to the skydiver, that produces a desire to repeat the experience' (Lipscombe, 1999, p. 281).

The most popular form of competition is *relative work* (also known as formation sky diving), during which groups of jumpers in free-fall exe-cute predetermined formations. This activity requires precise control of body position as two or more people direct themselves through various manoeuvres, including altering position to vary their rate of descent, performing turns and forward and backward movement. Other disci-plines are *accuracy landing, freestyle* and *canopy contact* or *stacking* (Carey, 1994; Crawford, 1996).

Base jumping and wingsuit flying are NCAs that may be undertaken by highly experienced sky divers. BASE is an acronym for the types of launching sites used for a base jump: building, antennae, span (bridges or any alternative spanning structure) and earth (e.g. cliffs, mountains, canyons). Jumping from lower altitudes than sky divers, base jumpers do not achieve the same airspeeds, which makes body stability and clean deployment of the canopy more difficult. In addition, with much shorter descents there is less room for error, as chutes must be deployed more quickly. An added challenge for a base jumper is to avoid colliding with the form from which they have jumped, either because they have failed to launch themselves adequately or because wind conditions have altered their trajectory. For these reasons base jumping is seen as being far more 'edgy' than sky diving, and sometimes deviant, as it is illegal to jump from many built structures (Ferrell et al., 2001). Wingsuits may be worn by either sky divers or base jumpers. These suits essentially shape the body of the wearer into an airfoil, with fabric between the legs and under the arms, allowing him or her to glide through the air. By this means, a jumper can prolong free-fall and cover extended horizontal distances before the canopy must be deployed. The specific challenge of the wingsuit is that it restricts physical movement making exits and deployments more difficult.

The first parachuting clubs were established around the world in the years following the Second World War and participation grew through-out the 1960s and 1970s. There are now hundreds of clubs in the United States, as well as Europe, South America, Asia and Australasia (Crawford, 1996). Membership of the US Parachute Association stood at 32,177 at the end of 2009 (www.uspa.org/Portals/0/Membership Surveys/memsurvey09.pdf, retrieved 18 March 2010), while there were 7200 sky divers in the United Kingdom participating on a monthly basis in 2007/08 (Sport England, 2009) and membership of the British Parachute Association is currently on the rise (http://www.bpa.org.uk/member, retrieved 18 March 2010).

A *drop zone* (DZ) is the area around and above a location where parachutists are jumping and intending to land. This may be beside a small airport where facilities are shared with other aviation activities. *Boogies* are local, national or international events held for 'jumping and partying' (Laurendeau & Sharara, 2008, pp. 30–31). An International Skydiving Licence is required to allow holders to jump at most drop zones throughout the world. Most national associations are affiliated with the FAI and are responsible for awarding the licences.

Parachuting is not an inexpensive hobby. A full sky diving system can cost several thousand dollars, depending on the brand and how new it is. The cost of training varies according to the type of opera-tion, and once divers have graduated from student status, and own and pack their own gear, there is still the cost of getting to altitude for each individual jump.

Paragliding

The paraglider, like the hang glider discussed in the following section, is a form of foot-launchable soaring aircraft. Paragliding is, arguably, the simplest and most intuitive form of human flight and therefore a powerful way to feel at one with the natural element of air. The wing is exceptionally light and portable, fitting into a backpack in which it can be carried up hillsides. Being independent of the need for motorized launching assistance, it allows a 'tremendous feeling of freedom... [and a] satisfyingly economical way to take to the air' (Carey, 1994). While it shares this characteristic – as well as similar training, theory, rules and regulations – with hang gliding, the paraglider travels more slowly and is therefore easier and safer to fly.

The evolution of paragliding can be traced through a series of advances in parachute and flexible wing technology and their recreational adaptation in the 1960s and 1970s. Critical to this was the work of Domina Jalbert, who had a life-long passion for kite flying and a professional background designing kites and balloons. In the 1950s he turned his attention to the parachute, with the objective of designing a canopy with directional control and a glide ratio of three to one (http://www.parafoils.com/jalbert/leo.htm, retrieved 19 March 2010). Jalbert's breakthrough came when he realized that glide would be best achieved by a fabric canopy that mimicked the shape of a wing. This led to his development of the parafoil, or ram-air wing, for which he was granted a patent in 1966. The rectangular ram-air wing is held in shape by air pressure inflating a series of cells incorporated within its design which are open at the leading edge and closed at the back (Carey, 1994). The cells are sewn together lengthwise, and when inflated each cell forms the cross section of an airfoil, together creating a semi-rigid wing-shaped canopy. With the improved manoeuvrability and glide ratio of the ram-air wing it became possible for enthusiasts to soar and glide under canopy. In the late 1970s, three French parachutists successfully inflated their ram-air canopies by running down the slope of a hill, and thereby launched themselves in the air. This innovation led to the growing popularity of paragliding in the French Alps.

A paraglider pilot is suspended below the canopy by risers connected to a harness and manoeuvres using control lines or 'brakes' which are attached to the trailing edge of the wing on each side. Pulling on either line will increase the drag on that side and cause the wing to turn in that direction. *Weight-shifting* can also be used to steer. Pulling on both brakes at the same time slows the canopy down. For a fast descent, both wingtips can be folded under using a technique known as 'big ears'. A foot-controlled 'speed bar', connected to the wing's leading edge, decreases the glider's angle of attack and increases the speed. Typically paraglider pilots will carry a reserve chute, variometer, radio and, if flying cross-country, a GPS.

Langewiesche (1998, p. 12) describes a paraglider launch:

> High on some mountain, you invert the fabric on the ground behind you, strap yourself into a seat-harness, and with a tug on the shrouds allow the wind to send the wing aloft directly overhead, where it assumes a cambered form and floats at the ready. With a short run downhill you give it flying speed. It answers by lifting you off your feet, and beginning to coast downhill toward the valley below. Once it gains speed it flattens its glide angle, and takes you out across the trees, the ravines, and the valley itself. The experience is primordial, a feeling of lift and wind like a throwback to the earliest elemental era of flight, before the Wright brothers, when pioneers like the great Berliner Otto Lilienthal floated downhill on homemade wings.

A paraglider gains height after launching by exploiting rising air currents, as Pagen (1996, pp. 407–408) explains:

> When wind strikes hills, ridges, and mountains, it is deflected upward, creating lift. Also, warm bubbles or columns of air known as thermals arise from the ground and often climb to great heights.... Finding the best lift in a thermal involves centering a series of circles on the thermal core, or fastest-rising current. Such a feat requires a three-dimensional sense of positioning and efficient turning technique. The problem is often complicated by the fact that thermals drift with the wind and often shift their core.

Thermals occur where the sun has heated geographical features to higher temperatures than surrounding surface features, thereby causing these locations to heat the air above them. Pilots develop the skill of finding thermals through their knowledge of the land features likely to generate them (such as rocky terrain, ploughed fields and asphalt) or by observing the cloud types that form at the top of them. A skilful pilot can travel cross-country by ascending in a thermal and then gliding to the next thermal to regain height and so on. A sound knowledge of meteorological conditions also enables pilots to avoid turbulent air, which can cause part of or the entire wing to deflate – although recreational wings will normally re-inflate themselves without the pilot having to act. Another hazard is high winds and wind gusts which can make take-offs and landings difficult. As well as flying technique and meteorology, a paraglider must learn aviation law and flight area etiquette. No pilot's licence is required for recreational paragliding, but a series of certificates and ratings are typically awarded by national organizations.

Paragliders talk of their experiences in terms of a peaceful solitude, a sense of freedom in the natural world and an 'incredible sensation of floating like a bird' (Moir, 1994). For obvious reasons, paragliding is more suited to mountainous regions, although it can still be done in flatter areas with a tow launch. While participants are now found around the world, it remains most popular in Europe where it first appeared, particularly in France where there were around 20,000 paragliders in 2009 (http://federation.ffvl.fr/sites/ffvl.fr/files/Plaquette%20statistiques%202009%20-%20def.pdf, retrieved 10 April 2010). The United Kingdom's monthly participation in paragliding in 2007/08 was 6200 (Sport England, 2009) and the United States has around 4500 pilots (http://www.ushpa.aero/safety/PG2008AccidentSummary.pdf, retrieved 10 April 2010). Estimates for other countries – from Cuba to Albania, China, Indonesia and the United Arab Emirates – have been posted on the site www.paraglidingforum.com (see http://paraglidingforum.com/viewtopic.php?p=87971).

Paragliders are one of the cheapest means of getting airborne. A mid-range canopy and harness will cost several thousand dollars and, depending on use, will last for about 200 to 300 hours of airtime. Training to the level at which you can fly your own canopy in a club environment costs around £900 in the United Kingdom.

Hang gliding

Early flying wings, similar in principle to a hang glider, were developed and flown in the late nineteenth century by the German engineer, Otto Lilienthal. These simple designs contributed to the development of powered flight, after which they were largely forgotten (Pagen, 1996). Then, in the 1940s, American aeronautical engineer Francis Rogallo, working with his wife Gertrude, developed a flexible wing in his quest to create a simple and cheap aircraft (*The Times*, 2009). While NASA explored the possibility of using the wing for the recovery of space capsules, the Rogallos' design was also exploited for its recreational potential and the activity caught on in the 1960s and 1970s throughout the United States and then Europe (Pagen, 1996).

Like paragliding, hang gliding is an affordable solution for people who want to fly as the equipment needed is minimal and can be transported in a car or a truck. The hang glider's triangular-shaped airfoil or 'flexible wing' is made of nylon or Dacron fabric supported by aluminium or composite tubes, and can be dismantled for transportation and storage. Pilots are suspended from the wing by a 'hang strap' attached to a harness that supports them, most often in a prone position, which helps to reduce drag. In front of them they grip the base of a triangular control frame, which is rigidly connected to the wing. Unlike a conventional aeroplane, a hang glider does not have any movable surfaces, such as a rudder or ailerons, which can be adjusted to control speed and direction. Instead, control is achieved by moving the whole wing, which the pilot does by pushing on the control frame and thereby altering the centre of gravity of the glider and the *angle of attack* of its wing (Goyer, 2004). Pushing forward provides more lift (and slows the glider), while pulling back causes the glider to descend more steeply (and gain speed). Similar movements to the left and right will lower the wing opposing the direction of the movement and bring about a turn into the lower wing. It is also possible to get rigid wing hang gliders that have spoilers for flight control. They tend to perform better than flexi wings, but are heavier, more complex and costlier (Pagen, 1996).

Hang gliders can be foot launched by running down a slope which has a gradient of six to one or more and faces into the wind. Other launch techniques include ground-based towing systems, boat or aero-towing, as well as powered harnesses. Once in the air, a hang glider finds lift in the same way as paragliders; that is, from thermals as well as air deflected upwards by landscape features such as ridges, mountains and dunes, and the convergence of air masses. Hang gliders have a higher glide ratio (horizontal distance covered in relation to altitude lost) and a lower sink rate (rate of descent in still air) compared to paragliders, which means they can generally stay in the air longer and cover greater distances. They can also reach higher speeds and their semi-rigid wings make them less susceptible to turbulence than the frame-less paraglider. At the same time, the flex in their wings means a gentler ride than an equivalently sized fully rigid-winged aircraft. However, the paraglider's superior turn radius can allow it to circle closer to a thermal core and with its slower speed, the paraglider is easier to land and requires a smaller landing zone (http://www.hanggliding.org/wiki/A_Comparison_of_Hang_Gliding_to_Paragliding, retrieved 26 March 2010).

High accident rates in the early days of hang gliding have since been greatly reduced by pilot training and improved design and construction (Goyer, 2004). In terms of safety equipment, pilots wear helmets and carry knives and a light rope (for extricating themselves in the event of tree or water landings), radios, first aid equipment and sometimes a reserve parachute. Instruments used include a variometer and altimeter, while more advanced pilots may also have airspeed indicators. Competition and cross-country pilots carry maps and/or GPS systems.

The first competition pilots tried to stay in flight for as long as possible and to make spot landings, but with the increased performance of hang gliders this has been replaced by cross-country flying where competitors pass waypoints to land at a goal or round a turn point and return (Pagen, 1996). Aerobatics are also performed. Arguably, the sensation of flying a hang glider is closer to flying like a bird than any other aircraft (Goyer, 2004). Depending on the flight conditions, the experience can be relaxing or can require intense physical exertion to control the wing. Kevin Frost describes how he responds to being asked what hang gliding is like:

> ...I go on and on about coring that big thermal and feeling the power and violent energy rocketing your glider into the sky, the wind roaring in your ears, and watching the mountains drop away into a spectacular panorama as you thermal up just a little short of airline cruising altitude where the temperature is literally bone chilling. If I have a captive audience I tell of a beauty few will ever experience – close-up views of the most rugged, inaccessible mountain peaks, the thrill of silently flying hundreds of feet from small waterfalls gushing out of rock faces invisible from anywhere normally thought of as accessible by foot. I strain to describe the thrill of seeing wildlife in a hidden meadow, a black bear and her cub traversing a shale slope or the surprise of seeing an eagle flying at your wingtip and looking you in the eye as you suck oxygen at 16,000 feet.
>
> (http://66.49.171.40/impress.htm, retrieved 27 March 2010)

Both paragliding and hang gliding are represented within the FAI by the Commission Internationale de Vol Libre (CIVL), which has delegates from all over North and South America, Europe, Asia and Australasia, but few African and Middle Eastern countries. In 2005 there were an estimated 7000 hang gliders and paragliders in the United Kingdom, including powered versions (see below), representing 26 per cent of all civil aircraft (http://www.caa.co.uk/docs/1/strateg%20review.pdf, retrieved 11 April 2010).

Hang gliders are more expensive than paragliders, but they are more durable and long-lasting. Like paragliding, it is a self-regulated sport and no licence is required. However, certification by a national body such as the United States Hang Gliding and Paragliding Association (USHPA) is often required to use dedicated facilities such as launch sites and landing zones.

Gliding or soaring

The British scientist Sir George Cayley is credited with designing and building the first heavier-than-air passenger-carrying glider in the mid-nineteenth century (Carey, 1994; W. D. Miller, 1996). From the 1890s onwards further developments were made towards the achievement of unpowered flights most prominently in Germany, England and the United States (W. D. Miller, 1996), but it wasn't until it was discovered that gliders could do more than gently descend – that is, that they were capable of soaring to considerable heights and thereby staying aloft for long periods and covering substantial distances – that the recreational potential of gliding truly became apparent (Carey, 1994).

Developments were halted by the First World War, but the sport grew throughout the 1920s and 1930s with Germany leading the way, prompted by the requirement of the Treaty of Versailles that prohibited the country from participating in powered flight. Gliding received new impetus again after the Second World War when many trained pilots wished to continue flying (W. D. Miller, 1996).

The terms 'glider' and 'sailplane' are often used interchangeably. Technically, any aircraft without an engine is a glider, while a sailplane is designed to achieve long soaring flights (Piggott, 1997, 2000). Gliders are designed to be as light and aerodynamic as possible, delivering maximum lift for minimum drag. The wings are much longer and thinner than a conventional powered aircraft, making them more efficient, and the fuselage is sleek and narrow, reducing drag and allowing the glider to slip through the air as easily as possible. The glide ratio for modern sailplanes can range from 20:1 to 60:1, according to their design and condition (W. D. Miller, 1996, p. 940), and some gliders are capable of loops, rolls and other aerobatic manoeuvres.

The common method of launching a glider is to tow it behind a powered aircraft by a steel cable, but skill is required to fly safely and smoothly behind the tow-plane. Alternatively, they can be launched from the ground via a car or winch tow, which is a simple and economical way of launching up to a height of 300 m (Carey, 1994).

A glider is manoeuvred in flight by a control column. When this is moved forward the nose of the glider drops, increasing the airspeed and the glider's rate of descent. If the column is moved back the opposite happens and the airspeed and rate of descent are decreased. Moving the column to the right, the right wing will drop down and the glider will bank to the right. These movements are the opposite of hang gliders and flexwing microlights described below, which can make it a challenge for pilots switching from one to another. A glider also has rudder pedals which can move its nose to the left or the right, and whose main use is to help balance the glider when it is turning. These basic movements, or *primary effects of control,* are the same for nearly all fixed-wing aircraft (Carey, 1994). Indeed, it is argued that gliding provides excellent experience for improving the skills of pilots of powered aircraft, as it teaches them to '*fly the wing* rather than the engine' (Goyer, 2004). Carey (1994, p. 80) describes the 'vivid sensations' of learning to fly a glider:

> The air is rarely still, and while you are flying you will often be tipped slightly this way and that by the currents in the air. While you are getting used to the glider and its controls these upsets can be rather disconcerting, but you will eventually correct them with the controls smoothly and automatically, rather as you react to traffic when driving your car.

Pioneer New Zealand glider pilot Dick Georgeson describes an early flight that helped cement his life-long passion for gliding:

> The little glider started rising quite quickly: with the green ball stuck at the top of the tube we were rising up the face of the mountain at over 1000 feet a minute.... it was as though the wings sprouted from my own back itself. It was a brilliant morning and the whole expanse of the Mackenzie Basin was unfolding before my eyes: Mt Cook stood out in splendour above the cloud that was pouring over the Copland Saddle and around Mt Sefton. There were the lakes beneath me and further south Lake Ohau sparkled in the distance. Curving above me was the limitless blue.
>
> This was life. This was living.
>
> (Georgeson & Wilson, 2003, p. 45)

Like paragliders and hang gliders, gliders must find and exploit rising air currents in order to stay aloft (W. D. Miller, 1996, p. 940). In locating thermals and other forms of lift, pilots are assisted by their variometers which indicate any changes in the rate of ascent or descent. Once a thermal is identified, pilots will attempt to fly in tight circles close to the centre of the thermal where it rises the fastest. By moving from one thermal to another, gaining lift from each, it is possible for sailplanes to achieve great distances, altitudes and durations. In winter, the sun's rays can only produce weak thermals, but ridge and wave lift can still be used. Strong ridge lift created by fast-moving weather fronts can support flights of over 1000 km (W. D. Miller, 1996, p. 942).

Mountain waves are produced when strong winds strike a mountain or ridge at a perpendicular angle before flowing over the top and down the other side, creating a 'wave' (or potentially successive waves) as they bounce first off a layer of stable air near the ground and then more stable air at higher altitudes. This meteorological phenomenon can provide very strong lift on the upward side of the wave, capable of carrying sailplanes thousands of metres into the air where special equipment including heated clothing and supplementary oxygen is required (Carey, 1994; W. D. Miller, 1996, p. 942). Flying wave requires experience and skill, particularly to avoid the turbulent 'rotors' that they create, but it is considered by some to be the ultimate of soaring flight, as Dick Georgeson explains:

> The experience of sailplane flying is incomplete unless it includes mountain flying, which has a peculiar thrill all of its own, transcending nearly all other aspects of the sport....to soar above the mountain peaks themselves, and look down on them spread out below, is an enthralling experience: a privilege for which one is forever grateful.
>
> (Georgeson & Wilson, 2003, pp. 76–78)

It is generally the intention of cross-country glider pilots to complete their flight at an appropriate airfield or other gliding site. If however they run out of lift they may be forced to 'land out' on a suitable field or other flat, unobstructed area. They must then contact their base to organize a retrieval. Alternatively, there are gliders that are fitted with small power units, either mounted in the nose or in retracting mounts above the fuselage, which allow for self-launching and save pilots from having to land out if they run out of lift (although this powered 'back-up' may detract from the satisfaction of 'pure' gliding for some).

At competitions, pilots compete on distance and speed tasks within defined areas, and aerobatic contests are also held. Soaring performance is recognized for gain of height, absolute altitude, duration, distance and speed (http://www.fai.org/gliding/system/files/sc3_2007. pdf, retrieved 1 April 2010).

In the United States glider pilot licences are issued by the Federal Aviation Administration (FAA), while in other countries, such as the United Kingdom and New Zealand, no formal licence is required to fly a glider, but a pilot's progress can be recognized by national awards and certificates for achievements such as solo flight, landing accuracy and other skills. Also available are the internationally recognized Silver, Gold and Diamond Badges each involving individual, documented and certified accomplishment in height attained, distance flown and duration achieved.

Most gliding participation occurs within clubs, as a team of willing helpers is required to get a glider into the air – unloading them from trailers, rigging them, operating the tow winch or providing the aerotow, and holding the glider steady before launching (Carey, 1994). Germany is still the world's centre for gliding, being the home country of just under 30 per cent of the world's glider pilots, followed by the United States with about 18 per cent and France with around 10 per cent. Monthly participation figures in the United Kingdom in 2008/09 stood at 8100 (Sport England, 2009), while world membership of FAI's International Gliding Commission was 111,492 at the end of 2006. There has been concern, however, about falling membership over the previous decade and failure to attract young people to the sport (www.fai.org/gliding/system/files/world_membership_2006.pdf, retrieved 30 March 2010). It has been suggested that the high cost of the sport discourages participation, particularly the need to pay for an aerotow (Goyer, 2004).

New sailplanes are expensive, costing tens of thousands of dollars, but can last several decades and there is an active second-hand market. Another popular solution is for syndicates of pilots to buy and share aircraft, and kits for homebuilding are available.

Microlighting

Microlights, or ultralights as they are also known, are lightweight aircraft that can be flown at low speed. The FAI defines a microlight as a one- or two-seat-powered aircraft whose minimum speed at Maximum Take Off Weight (MTOW) is less than 65 km/h, and having an MTOW of 300 kg for a landplane (flown solo) or 450 kg (two persons), with slightly greater weight allowances for amphibian and seaplane aircraft (http://www.fai.org/microlight/system/files/sc10_2010.pdf, retrieved 2 April 2010). There are two basic categories of microlight: the *flexwing* or 'weight-shift' and the *fixed-wing* or 'three-axis'. Other types include parawings and autogyros.

The flexwing has developed from earlier types of powered hang gliders and may be foot-launched or have a wheeled 'trike' unit beneath it. While still on the ground a trike is steered with the feet like a tricycle, with a throttle operated by the right foot and a brake on the left. Once in the air, the trike is controlled in the same manner as a hang glider, with a horizontal control bar used to shift the weight of the trike/pilot in relation to the wing (Finnigan, 2001; Goyer, 2004). A hand throttle controls speed, and thus the rate of climbing or descending. The appeal of a trike is that it is simple and relaxing to fly compared to the more sophisticated types of microlights and 'the direct connection with the controls makes you feel in closer contact with the air than any other form of powered flying' (http://www.bmaa.org/files/why_fly.pdf, retrieved 4 April 2010). Another advantage is their portability as they can be transported on a trailer towed by a family car (Carey, 1994, p. 59).

Fixed-wing microlights have evolved from efforts to make conventional aeroplanes lighter and simpler. They come in a wide variety of designs and are similar to aeroplanes in terms of the flight controls and experience, with the exception of being slower, lighter and more susceptible to turbulence (Carey, 1994, p. 59). They are also known as three-axis microlights because, like gliders and aeroplanes, they can be controlled along the three axes of pitch (aircraft pointing up or down), roll (raising/lowering the wing tips) and yaw (aircraft pointing left or right), whereas flexwings have only two axes of control – pitch and roll (http://www.bmaa.org/files/why_fly.pdf, retrieved 4 April 2010).

Early models were constructed from aluminium tube frames covered with fabric and were often powered by two-stroke engines. More recent, sophisticated models are made from carbon fibre and have purpose built engines, plus a full range of flight instruments in an enclosed cockpit. This generation of fixed-wing microlights is capable of high-speed flight over long distances and can withstand fairly rough conditions (Bain, 2001). The enclosed cockpit makes for a more comfortable experience, but loses some of the immediacy of the open flexwings. Likewise, microlights that are flown with less dependence on flight instruments allow for a more 'seat of the pants' flying experience. Some fixed-wings can be folded and taken home on a trailer, but otherwise they must be stored in a hangar at an airfield or other purpose-built shelter (http://www.bmaa.org/files/why_fly.pdf, retrieved 4 April 2010).

The main in-air control is a joystick with the same primary effects of control as a glider, with the addition of a throttle. Because of their light weight, fixed-wing microlights need less distance to take off and land than other aeroplanes (typically about 100 m). Another significant factor in their ability to become airborne within seconds and climb rapidly despite having small engines and low airspeeds is the 'high lift/high drag' design of their wings. These have comparatively more curvature on the upper-side of the wing, allowing them to produce a high lifting force. The trade off of this design is that it prevents them from flying at high speed and creates drag which reduces the glide ratio for microlights to about half that of other aeroplanes. With this in mind, microlight pilots need to ensure that they fly at sufficient altitude so that, in the event of engine failure, a suitable landing area is within reach. When flying at low altitude, it is particularly prudent to ensure that the terrain below could facilitate a landing as without power they descend very steeply (http://www.ultraflight.com/thornburgh/ DifferencesBetweenUltralightsandGeneralAviationAirplanes.pdf, retrieved 2 April 2010).

Parawings are essentially powered paragliders. The foot-launched version is known as a paramotor, and involves the pilot wearing a motor in a frame on his or her back which drives a propeller in a wire cage. Along with the two control handles attached to the wing, a pilot holds a throttle connected to the engine. A paramotor can be launched on flat ground, in still air or light wind conditions, 'after a short run into the wind with the engine roaring' (Finnigan, 2001; Langewiesche, 1998, p. 16). The paraplane is a wheeled version of the parawing, similar to the trike.

Along with paramotors, powered hang gliders – with a propeller mounted on the rear of their metal frame – make up the category of foot-launched microlights (FLMs). FLMs do not require a pilot's licence to fly and are therefore the least regulated of any form of powered aviation. Their simplicity, portability and low running cost also make them a very attractive form of powered flight (http://www.bmaa.org/files/why_ fly.pdf, retrieved 4 April 2010). The autogyro, which can also be categorized as a microlight if it is within the weight and speed restrictions, is discussed below in the section on rotorcraft.

Microlight enthusiasts, many of whom also fly or have flown more sophisticated aircraft, are drawn to this activity for the 'simple' and 'pure' fun of flying that it offers. Indeed microlighting has been described as a 'rebirth' of 'the love of low-speed flight which the earliest aviators felt so keenly, but which was subsequently lost in the quest for military superiority and commercial practicality' (http://www.bmaa.org/ pwpcontrol.php?pwpID=2409, retrieved 2 April 2010). Further pleasure is derived from the ability to fly 'low and slow' and enjoy the 'aerial view' of the landscape below, as well as the 'wind in your face' feeling of the open cockpit versions which allow the passengers to experience the smells and changing air currents and temperatures around them (Finnigan, 2001; http://www.usua.org/faq.htm, retrieved 4 April 2010). One drawback is that they cannot be flown in poor weather, but they can cross short bodies of water and are used for touring holidays (Finnigan, 2001).

Much of the original appeal of microlights was that they could be built at home from a kitset, flown with minimal (if any) training and were not subject to any regulation. After subsequent crashes, regulations were introduced, including training requirements and safety rules. In most countries now, with the exception of FLMs, a pilot must have a licence or be in the process of training for one (Carey, 1994). Licensing, however, is simpler and easier than for general aviation light aircraft, making the process more accessible and less costly. In the United States, no licence is required to fly an ultralight/microlight. Instead the US ultralight flying community is self-regulated with various organizations establishing training guidelines and issuing pilot certificates. Flying proficiency can also be recognized by the attainment of FAI badges and diplomas.

Evidence from a number of countries points to the growing popularity of microlighting as an NCA. As of January 2010 there were 4375 microlights registered in the United Kingdom, representing a 25 per cent increase over the previous decade (and up from 1600 in 1984), making them the most numerous class of aircraft after aeroplanes and one-fifth of registered aircraft overall (http://www.caa.co.uk/docs/56/REG% 20TOTALS%20WG%20010110.pdf, retrieved 11 April 2010). The CAA attributes this growth to their comparative convenience and affordability attracting more people to recreational flying, as well as luring existing pilots away from conventional fixed-wing aircraft (http://www.caa.co. uk/docs/1/strateg%20review.pdf, retrieved 11 April 2010). Membership of Recreational Aviation Australia Incorporated also presents a picture of growth. As of February 2010 its membership stood at 9400, with over 700 new members joining in 2009. At the end of 2009 they had almost 3000 aircraft on their register. A declining number of these were home-built aircraft (43 per cent, down from 53 per cent in 2006), and a growing proportion were flexwings or parawings (15 per cent, up from 12 per cent in 2006) (http://www.raa.asn.au/notices.html#membership_ dec10, retrieved 4 April 2010).

It has been estimated that the cost of microlight aviation is roughly a third of that for conventional light aviation (Carey, 1994). Costs for training and licensing or certification are lower, in part because fewer flying hours are required to attain this. As they are economical on fuel, microlights are also less expensive to run and owners can maintain their own aircraft. The greatest financial investment is in purchasing an aircraft. The cheapest option is an FLM, although one of these can still cost several thousand dollars. A flexwing trike or paraplane is the next cheapest option. Fixed-wings are the most expensive, with state-of-the-art factory-built models getting into the tens of thousands of dollars. Buying second hand is a money-saving option, as is buying a kit and assembling it at home. Some enthusiasts even design and build their own aircraft. An emergency parachute system is an additional cost, but a recommended safety feature in a microlight. Local clubs often provide trial flights and pilot training in club aircraft for a minimal cost, as well as organizing 'fly-ins' where participants can socialize and take part in organized activities and competitions. National and international championships are held with tasks testing navigation, precision flying and fuel economy.

Aeroplanes (light aircraft)

Aeroplanes are fixed wing aerodynes with a means of propulsion. While they owed much to their predecessors, such as Cayley and Lilienthal, the Wright brothers' first successful flight on 17 December 1903 is widely recognized as the beginning of the era of 'sustained and controlled heavier-than-air powered flight' (http://www.fai.org/news_archives/fai/000295.asp, retrieved 11 April 2010).

Smaller aeroplanes used for recreational flying are referred to as light aircraft and the majority of these are propeller powered. They may be high or low winged – depending on where the wings are attached to the fuselage – or biplanes. The types of light aircraft generally favoured by recreational pilots include the mass-produced two- or four-seat single-piston-engined training and touring aircraft which started to appear in the late 1950s. Most of the aircraft in this category have a simple fixed-pitch propeller and non-retractable undercarriage, although more sophisticated types are available. Lighter twin-engined aircraft may also be used for recreational flying. Floatplanes land on water using pontoons or floats, seaplanes put their fuselage in the water and amphibians have retractable wheels so they can also land on runways.

Aeroplanes have the same *primary effects of control* as fixed-wing microlights, described above. The advantages of an aeroplane over a microlight is its greater speed and range, and that it can be flown in a wider range of weather conditions. With the advent of instrument flying, in particular, it has become possible for pilots to overcome the spatial disorientation which can overwhelm them when engulfed in clouds. It is only with the aid of an artificial horizon, or attitude indicator, which is constantly in level with the earth's surface, that pilots have been able to fly 'inside' the weather (Langewiesche, 1998).

Like all aeronauts, aeroplane pilots develop a special sense of kinship with the natural element of air, as Langewiesche (1998, p. 112) explains:

> ... there comes a point in a pilot's life when the sky feels like home. In my case it came after 4,000 flight hours, during a certain take-off on a bright winter morning in Lincoln, Nebraska, westbound to California. Once airborne I retracted the landing gear and rolled into an early left turn, and as I looked back at the leading edge of the wing slicing stiffly above the frozen prairie, I realized that no difference existed for me between the earth and the sky; it was as if with these wings I could now walk in the air.

Pilots must acquire a private licence and ratings for various types of flying. Training is still a key function of aero clubs, although there are also private air schools. National aero clubs operate pilot proficiency schemes, and organize events and competitions for flight skills such as navigation, landing, aerobatics and formation flying. FAI competitions include precision flying, rally flying (including navigation and visual reconnaissance), air racing and formula racing, which consists of laps around a closed circuit course (www.fai.org/general_aviation/system/files/sc02_2010.pdf, retrieved 6 April 2010).

Participation rates for recreational aeroplane pilots are difficult to establish, as available figures on pilot licences and aircraft registration do not distinguish between leisure and non-leisure flying. The United Kingdom's CAA had 4600 single piston-engined aircraft on its register in 2002, which are most likely used for recreational flying, personal transport, and flying training. Registration of this type of aircraft grew steadily through the 1980s, with a sharp rise in terms of flying hours in the late 1980s. Since then this category has been relatively static in terms of active aircraft and hours flown, but it still outnumbers all other categories of aircraft – although the microlights are fast catching up. The separate categories of light twin-engined (420), vintage/historic and amateur-built (2000) aircraft also likely represent some NCA participation (http://www.caa.co.uk/docs/1/strateg%20review.pdf, retrieved 11 April 2010). Prices for single engine aeroplanes start at around US$100,000.

Rotorcraft

Rotorcraft is a category of aircraft that uses rotor blades rather than wings to produce lift. They include helicopters, gyrogliders (or rotor kites) and autogyros (also known as gyrocopters or gyroplanes). Experiments in rotor-powered flight in the early twentieth century culminated in the designs of Russian-American Igor Sikorsky that became the first mass-produced helicopters in 1942 and sported the rotor configuration that most helicopters still use today (Goyer, 2004). Helicopters are the realization of the dream of vertical flight and are an extremely versatile means of travelling through the air, being able to take-off and land virtually anywhere, to stop, hover and instantly change direction in mid-air (Carey, 1994). This manoeuvrability, however, means added complexity in terms of a helicopter's controls, and the development of good coordination in order to master them.

A helicopter has a *cyclic control stick*, which tilts the rotor disc in the desired direction of travel – left or right for a banking turn, forward/backward to lower/raise the nose. The *collective pitch stick* changes the pitch of the main rotors and controls the helicopter's lift. On the end of the collective pitch control is a twist-grip throttle. When climbing a small amount of additional power is applied via the throttle to ensure that the rotors keep turning within the required range of rpms to maintain lift. Anti-torque pedals adjust the pitch angle on the tail-rotor blades, allowing the pilot to 'yaw' the helicopter left or right and compensate for the torque effect created by the rotation of the main rotors. The challenge of helicopter flying is to coordinate these four *primary controls* – often simultaneously – in order to ensure stable flight. Hovering, for example, requires constant correction via the controls to eliminate any horizontal or vertical movement. If the engine fails in a helicopter, it can be disengaged from the main rotor via a clutch. With the rotor blades turning freely a helicopter can make a controlled descending glide using autorotation, as would an autogyro (although not with the same ease).

Helicopters are an exhilarating and challenging means of flight, but they are also among the most expensive (Carey, 1994). This limits their accessibility for most recreational pilots and perhaps their greatest role in the realm of NCAs has been as a means of transport to remote areas for the enjoyment of activities such as climbing, skiing, hunting and fishing. Nonetheless, helicopter flying is pursued by some as a recreation as well as a competitive sport. FAI-recognized helicopter competitions include tests of navigation and precision flying skills, including slalom courses and accuracy load drops. The non-obligatory free-style event is intended to demonstrate the highest level of helicopter manoeuvrability and pilot control, and may be accompanied by smoke and/or music to enhance the visual effect (www.fai.org/rotorcraft/system/files/cig_champrules_2010.pdf, retrieved 6 April 2010). Another popular optional event is the Bottle Opening, whereby a standard bottle opener is attached to the skid of the helicopter and used to take the top off a bottle secured to the end of a two-metre-high pole. The 13th World Helicopter Championship was held in Germany in 2008, with 44 crews from ten different countries competing. The FAI also records Helicopter World Records in a range of categories including speed, distance and altitude (see http://records.fai.org/rotorcraft for more information).

While the costs of helicopter flying are prohibitive for many aspiring rotornauts, the autogyro (or gyro for short) has made rotor-powered flight much more accessible. Although autogyros were the first successful rotary winged aircraft to fly, their recreational potential was not recognized until the 1950s when the engineer Igor Bensen developed a gyroglider, to which he soon added an engine to become what he termed a 'gyrocopter'. A gyroglider has no engine and therefore must either be towed into the air or dropped from another aircraft to initiate gliding flight. In the powered version, an engine-driven propeller drives the gyro forward and the resulting air movement under the rotor blades turns them. So instead of having powered rotor blades like a helicopter, the gyro relies on *autorotation* to produce lift. Thus a gyro requires a short roll along a runway in order to take-off, with the exception of some models which are fitted with a pre-rotator. Gyros can, however, land within the space of a few metres. And, as with the winged microlights described above, if the engine in a gyro fails it can descend in a glide to a safe landing, provided a small area of unobstructed, relatively flat ground is at hand.

Like helicopters, gyros are highly manoeuvrable, will not stall or spin and, as they are less vulnerable to turbulence, can be flown in conditions unsuitable for other microlights. The controls of a gyro are not as complex as those of a helicopter. In the absence of powered rotor blades a tail rotor is not required and therefore the pedals do not have the same 'anti-torque' function as those in a helicopter and do not need to be coordinated with the stick control. Because autogyros require forward movement to turn the rotor blades and produce lift they cannot hover, except in strong winds.

Flying a gyro has been likened to riding a motorcycle in the air and early gyros were primarily used for short, pleasurable flights but recently

more advanced models have made cross-country flights possible (http://
www.kiwiflyer.co.nz/KiwiFlyer-Issue2-Fly-a-Gyro.pdf, retrieved 5 April
2010). As with microlights enthusiasts talk of the feeling of 'real flying'
and enjoy the immediacy of a frequently open cockpit and the views
offered by slower flight at lower altitudes.

Gyroplane pilots are relatively few in number, but small and enthu-
siastic clubs organize fly-ins, and provide training and introductory
flights. In 2010 there were 300 gyroplanes and 1400 helicopters registered in the United Kingdom, with helicopters and gyroplanes making
up 6 per cent of civil aircraft overall (although these figures include helicopters used for non-recreational purposes, as well as inactive aircraft).
According to a review carried out by the CAA in 2006 there has been
a growing trend towards the recreational use of helicopters, facilitated
by the development of the more affordable light piston-engined two-
to four-seat helicopters such as the Robinson R22/R44 (http://www.caa.
co.uk/docs/1/strateg%20review.pdf, retrieved 11 April 2010). However,
2010 prices for these still start at US$250,000. The costs of gyroplanes are
comparable to those of microlights, and similarly they can be purchased
either new or second-hand, or built from plans or kits.

Amateur astronomy

As we will observe for all the amateur sciences covered in this book,
they began as amateur pursuits. More precisely stated theoretically: they
started as hobbies, since there was no professional counterpart with
which to form a P-A-P system. Interest in astronomical phenomena
dates to prehistoric times, as seen in, for example, Stonehenge, the
Egyptian pyramids and the Babylonia star catalogues of Mesopotamia.
Since then, and possibly even before, astronomical study was evident in China, India, MesoAmerica, and later, in Greece. An Islamic
astronomy emerged in medieval times. For the most part the rise of
European astronomy had to wait out the intellectual stagnation of
the middle ages. One common thread throughout these thousands
of years was that this science advanced entirely by the efforts of
hobbyists.

How is it that we can classify astronomy as an NCA? After all participants are not in nature in the same way as are, for example, parachuters
and airplane pilots. But, with the advent of the telescope (early seventeenth century), celestial observers could in effect be in nature, projected
out into the heavens, into nature, using powerful lenses and mirrors to
discover, describe and measure what they were beholding. Before that

invention, however, these observers were merely onlookers, watching stars and the like much as spectators might watch people hang gliding or sky diving. No natural challenge in that.

Historically astronomers busied themselves with classification and description of heavenly phenomena, and this remains the principal domain of the modern amateur (formerly hobbyist) in this field. In practice this means that the vast majority of amateurs work with telescopes, either ones they have built themselves or ones purchased at some considerable cost. Stebbins (1980) describes their leisure routine in this activity and the many challenges to be met using a telescope to find, describe and measure stars, comets, meteors and the like. In contrast their professional counterparts do much less observational work (describing/classifying) and much more theoretic work. In this capacity they are basically astrophysicists.

Rothenberg (1997) has chronicled the development of professional astronomy in the West. He observed that:

> as paid positions for astronomers gradually became more common early in the nineteenth century, the profession of astronomy began to emerge, and with it, the need to distinguish those who were amateur astronomers. An amateur astronomer is one who practices astronomy as a science but without pay, a meaningless distinction before about 1800. (pp. 6–7)

Thus between 1800 and 1860 amateurs were increasingly distinguished from professionals, and by approximately 1920, it had become nearly impossible for an amateur to make the transition to professional status without full academic credentials.

Timothy Ferris (2002, p. xvi) discusses in the Preface of his book the wonder amateur astronomers feel:

> Although this book is not intended to be a how-to manual of amateur astronomy, I hope that it will encourage its readers to make the glories of the night sky a part of their lives. The universe is accessible to all, and can inform one's existence with a sense of beauty, reason, and awe as enriching as anything to be found in music, art, or poetry.

Why do serious amateur stargazers stick with their passion?

> Some mention the beauty of the planets, stars, nebulae, and galaxies. Others invoke the grandeur of the cosmos and their sense of

belonging to it. A few mention that contemplation of the stars draws people close to one another awakening us to our common status as fellow travelers on one small planet.

Amateur meteorology

Aristotle is credited as being the founder of meteorology, although observation of the weather was not something new in the fourth century B.C. By the eighteenth century, amateur scientists throughout Europe and North America were recording weather conditions, providing the foundations for the development of observer networks and important advances in the science during the nineteenth century (Fine, 2007; Littin, 1990). One such breakthrough was the description and naming of the various cloud types in 1802 by the amateur English meteorologist, Luke Howard. The elegance and mutability of Howard's system, according to Hamblyn (2001, pp. 180, 251), 'reaffirmed meteorology as a science of contemplation' and 'completely transformed the relationship between the world and its overarching sky... "The ocean of air in which we live and move, with its continents and islands of cloud," he once wrote, "can never to the conscious mind be an object of unfeeling contemplation".'

Most NCA participants require some appreciation of meteorological phenomena. Pilots and long-distance sailors in particular need to have a sophisticated grasp of the science of the weather and the skills for accurate forecasting. For others, meteorology is sufficiently challenging and inspiring as a core activity in its own right. Here the challenge is in tracking storm patterns, temperature and rainfall, understanding how climate changes over time and successfully predicting future weather events. Participants must be able to read weather maps, comprehend meteorological terminology and the principles and characteristics of various weather patterns. While most of us yearn for fine sunny days, it is the dramatic aspects of the weather that excite the amateur meteorologist – tornados, lightning, bizarre cloud formations. As the editor of *Weatherwise* – a journal devoted to 'the power, beauty, and excitement of weather' – notes, 'unusual weather is never bad news' for meteorologists (http://www.weatherwise.org/Archives/Back% 20Issues/2010/March-April%202010/from-editor-full.html, retrieved 23 April 2010). The journal's readers are kept abreast of meteorological news and events, with 'expert columnists' writing on current issues and advances, and testing 'reader forecast skills with analysis of weather maps' (http://www.weatherwise.org/About%20Us/about weatherwise.html, retrieved 21 April 2010).

Amateur meteorologists strive to produce accurate weather forecasts, with the ultimate challenge being to better the official forecasts, and take pride in being able to provide people with useful information. Herb Hilgenberg, for example, was inspired by personal experience to provide more accurate forecasting for the marine environment. He acquired a Ham radio licence so that he could gather information from ships at sea and began studying weather maps. His reputation for forecasting grew and sailors began calling him up for his 'on air' weather advice (Cutlip, 1999).

For storm chasers, a peculiar subset of amateur meteorologists, forecasting is the means to facilitate a chance to personally witness some of the more awe-inspiring forces of the weather. Svenvold (2006, pp. 7–8) explains the attractions of one of the most intriguing and challenging-to-predict meteorological phenomena:

> Every tornado represents a supreme, if momentary, trouncing of the second law of thermodynamics, the glum law that all things move from order to chaos. Tornados move the other way, from a chaos of cloud swirl, from a mixture of lines of force, density, temperature, lift, speed, and convergence, a set of initiating conditions whose exact ingredients are still unknown, to a near perfect level of order and organization, capable, paradoxically of delivering immense destruction ... [The storm chasers] came for the same reasons people choose to encounter other natural wonders ... they came for the wonder of it, for a bit of self-induced awe.

Most 'weather hounds' collect data about the weather using reasonably inexpensive instruments that can be easily purchased and installed in their backyards. These include a weather thermometer, a barometer (for atmospheric pressure), a hygrometer (humidity), a wind vane (wind direction), an anemometer (wind speed), a good compass, an atomic clock and a rain gauge. More sophisticated multi-component 'home weather station' packages can also be purchased for several hundred dollars, complete with wireless capability to send digital read-outs to a computer using specialized software. Top-end units have the capacity to allow for climate forecasting and prediction and provide severe weather alerts. Storm chasers, in their quest to be mobile, mount their equipment on the back of SUVs (Svenvold, 2006).

In addition to tracking their local observations, amateur meteorologists can access websites like the National Weather Service in the United States which has a variety of tools and maps including radar, satellite, air quality, climate and water, and weather-related forums and blogs are frequented by both professionals and amateurs. Computer technology has revolutionized the potential for weather forecasting using sophisticated models and huge databases of information, while the Internet has greatly facilitated the networking opportunities and a two-way information exchange between professionals and amateurs. A tradition of 'citizen weather observers' which began in the nineteenth century continues today (Littin, 1990), with extended networks linked together online.

In the United States, the Citizen Weather Observer Program (CWOP) (www.wxqa.com) invites people with home weather stations and Internet access (or a wireless digital Ham setup) and who 'think of watching weather as a serious hobby', to provide real-time surface observations. This data is sent to a server at the National Oceanic and Atmospheric Administration (NOAA) and then distributed to users. Volunteers monitor this information around the clock when hurricanes threaten. The CWOP has 8000 registered members worldwide. In the United Kingdom the Climatological Observers Link provides a national weather observers network, with a membership of more than 400 (http://www.met.rdg.ac.uk/~brugge/col.html, retrieved 17 April). Another US-based network is Weather Underground (www.wunderground.com) which claims to be the world's largest network of personal weather stations, with almost 10,000 stations in the United States and over 3000 across the rest of the world. Their website provides real-time online weather information, as well as blogs and other services aimed at promoting the community aspects of this NCA. Storm chasers frequent Internet forums such as www.stormtrack.org and www.ontariostorms.com.

Conclusions

The main sustainability issues for air sports are carbon emissions, noise pollution, disturbance of wildlife and the take over of open space by aircraft facilities. The FAI has an Environmental Commission dedicated to working on these issues and has developed a Code of Conduct. This promotes measures such as purchasing new aircraft with noise suppression, flying in economical ways and reducing energy consumption wherever possible, for example by using winch tows rather than aero tows. Other recommended measures include 'adhering to optimum climb and descent rates' and planning flight paths that cause the least noise and disturbance as possible; respecting wildlife and not flying low over sensitive public areas. Aircraft that land away from airports are directed to avoid doing so in such a way as may cause erosion or disturb the flora and fauna, including prepared fields and meadows.

Powered aircraft have been found to cause a range of reactions in birds, from increased heart rates and reduced food intake through to 'panic-like flight reactions'. In particular, helicopters can have detrimental effects on wildlife, including affecting habitat use and the raising of young, and are considered more disruptive than fixed-wing aircraft (http://www.wildlandscpr.org/node/218, retrieved 6 April 2010). The cumulative effects of disturbance by air traffic may result in the loss of territory for some bird species. Low flying balloons can also trigger reactions, and studies in Germany have recommended more detailed aeronautical maps indicating bird sanctuaries and better information and guidelines on how to avoid disturbing birds, particularly endangered species, such as a minimum fly-over altitude of 600 m (or 300 m for balloons) and areas where aircraft should not land (see www.fai.org/environment/wildlife for more information).

A positive trend in terms of the sustainability of air NCAs is the growth of what has been termed 'New Aviation'; that is, the lower cost, unpowered or low-powered activities of hang gliding, paragliding, microlighting and gyro flying (Finnigan, 2001, p. 17). However, even these are not without their impacts. Microlights, like balloons, can have a negative effect due to the low altitude at which they are flown. Meanwhile, predatory birds have been known to attack or 'pseudo-attack' gliders, hang gliders and paragliders during breeding season, and these aircraft may also induce greater anxiety in chamois goats and ibexes than other aircraft, including helicopters.

In terms of consumption, the air NCAs offer a range of options depending on one's disposable income. However, they do all involve a relatively high initial outlay for equipment and training, with the exception of amateur meteorology which could be pursued in a reasonably economical way (although with the increasing use of computers and the Internet the trend in this NCA is certainly towards higher consumption levels).

3
Water

Around 70 per cent of the Earth's surface is covered with water. It is no wonder then that humans have found various means of travelling over and under water for millenia, in order to get places, find food and wage wars. How much of this early engagement with the element of water was for playful purposes is hard to say – some of it was certainly competitive. But perhaps more importantly, while many of the original means and motivations for being on or in the water have become obsolete, *homo otiosus* has resurrected them, adapted and evolved their original forms into numerous variations for recreational ends.

Unlike the air NCAs then, most of the activities discussed in this chapter have ancient origins or, at the very least, combine age-old practices and technologies with recent innovations, such as the aerodynamic parafoil, the internal combustion engine and the aqualung. Water, for its part, provides a wide range of environments and conditions, in a state of constant flux, with which to engage. Whether it is salt water or fresh, above the surface or below, flowing, flat or white caps, inshore or offshore, every facet of water seems to have been exploited as a natural challenge. Neither does water disappoint in terms of its ability to inspire awe in its recreational enthusiasts: from the vastness of the ocean and coral reefs teeming with aquatic life to the clarity and serenity of remote lakes, and the thundering of river rapids through a gorge or canyon. The ambience of the aquatic environment varies between NCAs, allowing participants to feel a connection with nature in diverse ways. It might be floating in the silence and strangeness of the underwater world, experiencing the power of the surf on a board, or feeling the unity of self and natural forces while executing precise manoeuvres on a river rapid or – with the addition of man-made propulsion – performing an aerial trick on water skis.

We begin this chapter with open water swimming, the simplest of activities and one that all aquatic recreationists are likely to engage in – not always by choice – but which is also pursued for its own sake. This is followed by the underwater activities of snorkelling, freediving and scuba diving. The board sports are next – surfing and its close, but wind-drawn, counterparts windsurfing and kiteboarding. Staying with the theme of wind, sailing follows, before we progress to its fellow boating activities, with the self-propelled craft first – canoeing, kayaking, whitewater rafting and rowing – and then the motorized versions. We finish with the towed activities of water skiing and wakeboarding which put together the speed afforded by power boats with technologies and skills blended from skiing, surfing and snowboarding.

Open water swimming

Open water, or wild, swimming is a swim in any natural body of water. Of course all swimming was once open water, and early races were held in Ancient Rome and Japan (http://www.10kswim.com/history.html, retrieved 25 April 2010). A milestone in long-distance open water swimming was the first crossing of the English Channel in 1875, by Captain Matthew Webb. When the Olympic Games were reinstituted in the late nineteenth century, swimming races were held in open water and competitors had to contend with river currents and heavy surf. While these events retreated to the calm and predictability afforded by swimming pools, some interest remained in open water, with the first major international race being held across the Catalina channel in California in 1927 (Dean, 1998). Aside from competitive open water swimming and individual swims across challenging bodies of water, groups of recreational swimmers remain dedicated to swimming in oceans, rivers and lakes, rather than man-made environments.

According to former world record holder Penny Dean, the challenge of open water swimming is 'competing with the elements... No one can control Mother Nature' (Dean, 1998, p. 3). Cold water is particularly testing, as temperatures below 19 degrees centigrade can cause physiological problems and lead to hypothermia. Body fat provides some protection, and a bathing cap – brightly coloured silicone caps are recommended for both warmth and visibility – and ear plugs help prevent heat loss from the head. Some swimmers cover their bodies in grease to help retain heat, or wear wet suits, although 'non-thermal suits' are a requirement for many open water races and individual record attempts (Dean, 1998, p. 12). An alternative is a body suit, which is like a regular swimsuit, but covers more of the body. Beyond these measures, swimmers must acclimatize their bodies to the cold: entering the water slowly and gradually increasing the length of time spent there while remaining vigilant for signs of hypothermia.

Different bodies of water pose distinct challenges. River currents can be fast and rough, and on lakes wind can create a choppy surface, while fresh water feels more intensely cold than salt water. Ocean swims are affected by wind and currents too, but their speciality is waves and tides. In the English Channel, for example, adverse tides can add an extra 6 hours to a crossing and are one of the major reasons for unsuccessful attempts (Dean, 1998). Although rarely a serious threat, marine life such as jelly fish and sharks may cause swimmers anxiety, while in fresh water there are lake reeds and leeches to avoid.

Breaststroke was still being used prior to the 1950s, but freestyle is a faster and more efficient stroke, and is now the most popular for open water swimming. The style is slightly different from that used in a pool, with a focus on speed, efficiency and endurance (Dean, 1998). The arms are the main means of propulsion, with the leg kick primarily providing stabilization. Breathing patterns are regular, with breaths to alternate sides helping to maintain a symmetrical stroke and a straight course. In rough conditions the swimmer may have to breathe on the side opposite to the waves, and/or turn the head further than normal to ensure their mouth is above the water. Another important technique is *spotting*, or raising the head on a stroke to check one's course.

Individual swimmers on long swims are typically accompanied by a support craft which assists with navigation and provides them with food and water, as well as monitoring for signs of illness and pulling them from the water if they are in trouble. They can also keep a look out for marine life and shelter the swimmer from poor weather.

Competitive open water swimmers 'conquer the challenges of the ever-changing weather conditions, tides, currents, hypothermia, heat stroke, marine life, and the loneliness to realise their dreams' (Dean, 1998, p. vi). Their recreational counterparts, such as the members of the Outdoor Swimming Society in the United Kingdom, share a desire to escape the 'chlorined captivity' of swimming pools, 'embrace the rejuvenating effects of cold water' and 'immerse themselves in nature' so that they may 'recover a sense of how the natural world smells, tastes and sounds'. Their Patron, Robert Macfarlane enthuses about the sensory and aesthetic pleasures of open water swimming:

> To enter wild water is to cross a border. You pass the lake's edge, the sea's shore, the river's brink – and you break the surface of the water itself. In doing so, you move from one realm into another: a

realm of freedom, adventure, magic, and occasionally of danger....
Everything alters, including the colour of your skin: coin-bronze in
peaty water, soft green near chalk, blue over sand. You gain a stealth
and discretion quite unachievable on land – you can creep past chub
and roach, or over trout and pike, finning subtly to keep themselves
straight in the current.... And the smells! The green scent of the
riverbank. The estuary's Limpopo-gunge whiff. The mineral smell of
high mountain lakes.... you can never go for the same wild swim
twice. Weather, tide, current, temperature, company – all of these
shift between swims. And different types of water actually *feel* different. Wild water comes in flavours. Not just salt and fresh, but
different kinds of fresh.... Let's be clear, though, wild swimming
is about beauty and strangeness and transformation – but it's also
about companionship, fun, and a hot cup of tea or nip of whisky
afterwards.

> (http://www.outdoorswimmingsociety.com/index.php?
> p=about&s=, retrieved 21 April 2010)

As an international competitive sport, open water swimming is gov-
erned by the Fédération Internationale de Natation (FINA), which
represents 202 national federations. Since the 1990s FINA has organized
a global World Cup and the Grand Prix Open Water pro series and
according to their website the sport is growing and spreading around the
world as it becomes 'associated with racing the waterways of many beautiful and exotic destinations', from Manhattan to Copenhagen and the
United Arab Emirates (www.fina.org/project/index.php?option=com_
content&task=view&id=2440&Itemid=52, retrieved 29 April 2010).

Open water swims are also a component of triathlons, the huge popularity of which is further promoting this NCA as triathletes enter open
water races to improve their swimming (Dean, 1998). According to
the former open water champion and coach, Steve Munatones, both
triathlons and open water swimming races have experienced 'explosive'
growth over the past few decades. From his survey of swim events across
81 countries, Munatones found that the average number of entrants had
increased by almost 80 per cent in the past decade. Some popular events
have experienced exponential growth, such as the Midmar Mile race in
South Africa which began with 153 swimmers in 1974 and had 17,575 in
2009 (http://www.thewaterisopen.com/news/full/triathlons_and_open_
water_-_enduring_explosive_growth, retrieved 5 May 2010).

Numbers of recreational participants are harder to ascertain, as most
available statistics do not distinguish between indoor and outdoor

swimming. The River and Lake Swimming Association in the United Kingdom is made up of nine open water swimming clubs, with 600 individual members. The association's website notes that open water swimming was a popular leisure activity in the 1920s and 1930s, but is only recently showing signs of a revival after losing widespread appeal in the face of competing leisure options and the adverse effects of health and safety regulations (http://www.river-swimming.co.uk/about.htm, retrieved 21 April 2010).

Open water swimming is perhaps the least consumptive NCA of all. As one proponent puts it:

> What could be more democratic than swimming? What more equalising than near-nakedness? You need even less equipment to swim than you do to play football. A bathing costume, if you insist.
> (http://www.outdoorswimmingsociety.com/index.php?
> p=about&s=, retrieved 21 April 2010)

Snorkelling and freediving

A snorkel allows a swimmer to stay face down on the surface of the water, without having to lift the head to breathe. With the addition of a mask comes the pleasure of underwater viewing, and a pair of fins provides propulsion that increases one's range. The use of a snorkel and fins is simple to master and often the first step in learning to dive. Those content to stay on the surface must still take account of water and weather conditions, including currents and tides. To leave the surface requires additional skills: a duck dive, or similar, to begin the descent, finning techniques, using either bi-fins or a monofin, and purging the snorkel of water on regaining the surface in order to resume breathing (Graver, 1999). A weight belt can be used to assist with the descent, a buoyancy vest for the ascent and a repertoire of hand signals facilitates underwater communication.

To dive on a single breath is known as freediving, breath-hold diving or apnea. Its historical antecedents are the traditions of spearfishing and diving for shellfish, pearls and sponges. Today freediving generally involves the use of breathing techniques, efficient movements and specialized equipment to reach greater depths and spend prolonged periods of time underwater for sport or recreation. The development of fins and improvements in mask design in the 1930s were important for facilitating this recreational evolution. In the same decade, in the Mediterranean, hobbyist spearfishing was catching on (Ecott,

2001). In 1949, Hungarian-born Italian Raimondo Bucher, an experienced spearfisher, dove to 30 m on a breath-hold for a wager with a friend. With Bucher setting the standard, Italy became the centre of the fledgling sport of competitive freediving. In 1962 Enzo Maiorca, whose rivalry with the French freediver Jacques Mayol was depicted in the Luc Besson movie *The Big Blue* in 1988, broke the 50 m barrier which until then scientists had considered an impossibility due to the pressure exerted on the lungs at that depth. In 1976 Mayol reached 100 m using a weighted sled (Fleury, 2007). Besson's movie, with its representation of the Zen-like qualities of freediving, helped bring more global attention to the then still obscure activity. The first world championship was held in Nice in 1996 (http://www.aida-international.org, retrieved 6 May 2010).

Competitive apnea, with its physiology-defying feats of depth and endurance, is perhaps the most widely known aspect of this NCA. However, freediving is also practised as a recreational means of exploring the underwater environment. The core challenge for freedivers is to suspend breathing for as long as possible and successfully manage the physiological effects of a lack of oxygen and underwater submersion. This process is aided by the natural adaptations of the human body known as the 'mammalian diving reflex', including a decreased heart rate and other cardiovascular changes to help conserve oxygen. This reflex is triggered automatically if the face and nose are immersed in cold water, leading to speculation about our evolutionary connections to the aquatic environment (Farrell, 2006; Fleury, 2007).

Relaxation is the key to breath-hold diving, as any stress or tension will more quickly deplete precious oxygen reserves. While science has demonstrated that the early practice of hyperventilation is ineffective and even dangerous, divers such as Jacques Mayol have introduced techniques adapted from Eastern yoga and meditation traditions. These enable the freediver to take deep, full breaths using the diaphragm, sometimes inhaling extra gulps of air (a technique known as *lung packing*) to optimize lung capacity and the oxygen available during a dive. Resisting the urge to breath requires mental discipline as well as increasing the body's tolerance of high carbon dioxide levels in the blood, which trigger it. Training techniques known as CO_2 tables help develop this tolerance.

To avoid pain and potential ear injuries, freedivers use equalization techniques during their descent to increase the air pressure in the middle ear and thereby counter the effect of mounting water pressure (Fleury, 2007). Buddy systems and hand signals help to ensure safety, particularly if a diver suffers *black out* from critically low levels of oxygen.

A freediver describes the experience:

> You float on the surface, breathing deep and slowly, watching the bottom 40 feet below you... In tune with your body, you begin to slow your heart rate down.... You take big, almost stiff-legged kicks, very slow frequency, but with large amplitude. You almost try to sleep your way down to the bottom, staying as relaxed as possible, exerting the minimum amount possible. Twenty seconds later you are moving along the bottom about 50 feet deep, effortless and free, watching the antics of damselfish, triggers and the constant schools of mangrove snappers among the bright hues of sponges, gorgonians, and complex coral colonies. You glide along at close to twice the speed you could have traveled with scuba gear on, but with virtually no exertion.... You begin to imagine the efficiency of function a fish or dolphin must realize with their superior hydrodynamic designs...
>
> You tilt upwards... You stay relaxed, you maintain a slow steady kick, and the surface approaches.
>
> (http://www.sfdj.com/sand/freedive.html, retrieved 7 May 2010)

Like the aeronauts in the previous chapter who felt a kinship with birds, freedivers experience an affinity with marine life. Although they share this with scuba divers, freedivers maintain that while fish are easily startled by the noise of regulators and the trail of bubbles emitted by bulky scuba equipment, the relaxed and fluid movements of freedivers allow a closer empathy with aquatic animals. It also makes them all the easier to hunt and photograph, and hence the use of breath-hold diving for these activities (see Chapter 5). Another frequently expressed aspect of freediving is the serenity of the meditative state induced by submersion in the silent underwater world and the physiological effects of the breath-hold (Fleury, 2007). UK freediver Emma Farrell explains these sensations and what they mean for her:

> As a child I dreamt of weightlessness.... the freedom to move in three dimensions, would always elude me until I learned to slip beneath the surface of the sea and explore a liquid space... On one breath, under the water, our sense of smell and taste are stilled and our hearing muted. We are left with our sight and the invisible caress of the water

moving around our bodies. This narrowing of external stimuli opens the petals of our sixth sense, which flourishes in the void.

Our sixth sense reveals deep connections and feelings that have no logic and cannot be described with words. It takes us from the tiniest particles within us to the vastness of the universe beyond. We are all separate and at the same time we are all one.

... our breath is the most fundamental and constant expression of life, and when we connect with it fully, it vibrates with joy through every cell. Freediving connects us with our breath, our life and the world that nourishes us. It enables us to touch our history, our soul, and feel peace at a profound level.

(Farrell, 2006, pp. 91, 106)

The world governing body for freediving, l'Association Internationale pour le Développement de l'Apnée (AIDA), was established in 1992. AIDA organizes competitions and recognizes world records in the various competitive apnea disciplines. Some disciplines are performed exclusively in pools, while static apnea (where the competitor floats face down in the water and holds his or her breath as long as possible) can take place in a pool or open water. The open water events focus on descending to the greatest depth possible and returning to the surface, the purist of these being *constant weight without fins*, which requires 'perfect coordination between propulsing movements, equalization, technique and buoyancy'. Other disciplines use varying combinations of fins, rope pulling, ballast and inflation devices to move through the water (http://www.aida-international.org, retrieved 8 May 2010).

Competitive apnea and spearfishing are also represented at the international level by the Confédération Mondiale des Activités Subaquatique (CMAS). Overall participation rates are difficult to gauge. While memberships are currently small – the British Freediving Association for example had 132 members in October 2009 (http://www.british freediving.org/index.asp?sec=13&pag=137, retrieved 8 May 2010) – anecdotally participation is growing and new clubs are being established. In its favour, freediving is a relatively inexpensive means of entering the underwater world. While the start-up costs include the purchase of specialized equipment including a mask, snorkel, fins, wetsuit and weight belt, dispensing with artificial breathing apparatus is a considerable saving.

Snorkelling, however, is one of the more popular water-based NCAs, with around 10 million participants in the United States in 2008, or

3.7 per cent of the population (Outdoor Foundation, 2009a). In the United Kingdom 12,200 people went snorkelling at least once a month in 2008/09 (Sport England, 2009).[1]

Scuba diving

To prolong their visits underwater beyond the limits of one breath, divers use scuba (Self-Contained Underwater Breathing Apparatus) equipment, which allows them to inhale compressed gas, usually air, through a mouth piece.

Attempts to design a means of breathing underwater date back centuries. In the 1920s Frenchman Yves Le Prieur began developing a scuba device with an air cylinder that allowed him to swim freely underwater while breathing. Building on this technology, Jacques-Yves Cousteau and Emile Gagnan produced the 'Aqualung' in 1943 (Ecott, 2001). This open-circuit, free swimming underwater breathing apparatus is the predecessor of the majority of recreational scuba units used today.

A modern-day scuba diver is equipped with a mask, snorkel, fins, exposure suit (either a wet suit or a dry suit, to provide protection from scrapes and stings and provide insulation), a weighing system (either as a belt or integrated into the scuba unit), a buoyancy compensator (inflatable/deflatable vest), a scuba unit (cylinder, valve, regulator, alternate air source), instrumentation (depth gauge, underwater timer, dive compass, submersible pressure gauge to measure the tank pressure), dive knife (in case of entanglement) – and for cold water – a hood, boots and gloves (Graver, 1999). The scuba cylinder or tank contains compressed air at high pressure and has a valve to control the flow of a liquid or gas. The regulator reduces the compressed air to a suitable pressure for comfortable breathing and the alternate air source (AAS) provides a backup in case the primary source is exhausted or fails.

Scuba divers must master the use of their equipment while entering and exiting the water, descending and ascending, and moving about *sub aqua*. An important skill is controlling buoyancy which is affected by the amount of air in the lungs and the tank, and suit compression. Skilled scuba divers can hover horizontally and vertically, controlling their average lung volume while breathing continuously and achieving what is known as 'neutral buoyancy' (Graver, 1999). When ascending and descending divers manage buoyancy by inflating/deflating their buoyancy compensator.

Divers can enter the water by wading in from the shoreline, or making a seated, feet-first or back roll entry from a suitable platform or

the back of a boat. They descend feet first to best control their buoyancy, maintain their orientation and communicate with their buddies
(Graver, 1999). While descending they must equalize their ears as freedivers do. The ascent should be gradual, with a decompression stop for
several minutes a few metres below the surface to allow for 'outgassing';
that is, the elimination of nitrogen that builds up in the body at depth
(Graver, 1999). If the ascent is too fast – and therefore the reduction
in pressure too rapid – the nitrogen will form bubbles in the tissues
or blood, causing Decompression Sickness (DCS), otherwise known as
the bends, which in the worst cases can cause permanent disability and
death. The nitrogen build up also means that there is a limit to how long
a diver can remain at various depths. Time at depth and the intervals
between dives need to be carefully planned and calculated to avoid DCS.

Another concern is nitrogen narcosis, caused by breathing high pressure gas at depth. This can lead to an altered state of consciousness that
Jacques-Yves Cousteau famously referred to as the 'rapture of the deep'.
Despite the euphoria it may induce, narcosis can result in a potentially
hazardous loss of judgement, coordination and concentration. However, the symptoms normally subside on ascending. While underwater,
divers must monitor their instruments every few minutes to keep track
of depth, time, direction and tank pressure. They should stay close to
their diving buddies so that they can check each other's equipment,
observe and assist as necessary.

The rewards of scuba diving have many similarities with freediving:
the entering of a new world, filled with intriguing aquatic life and
unusual formations of rock and coral. Scuba divers, too, speak of the
joys of weightlessness and feeling 'the freedom of a bird as you move in
three dimensions in a fluid environment' (Graver, 1999, p. 2). The comfort and familiarity of the underwater world is sometimes connected to
memories of childhood desires, and is associated with feelings of peace,
solitude, mental calm and a connection with the ecosystem (Dimmock,
2009; Wynveen et al., 2010, p. 280).

Recreational scuba diving is mostly self-regulating, although dive centres and equipment suppliers require proof of diving competency in
the form of certification from one of the main certifying bodies, such
as the Professional Association of Diving Instructors (PADI). Most participation is recreational, but scuba divers can compete in underwater
orienteering, photography and spearfishing. Events such as the British
Sub Aqua Club's annual Dive Fest provide opportunities for scuba divers
to gather together. Specialized scuba activities include cave diving, wreck
diving, ice diving, altitude diving and deep diving.

It has been estimated that there are 5–7 million divers worldwide and that scuba participation has been growing in recent years (Dimmock, 2007). However, figures from the United States show around 1 per cent of the population or 2.7 million people went scuba diving in 2009, which represents a 36 per cent decline in participation since 2000 (Outdoor Foundation, 2010). Australia's participation rate was lower at 0.5 per cent, or 90,200 people in 2008 (Australian Sports Commission, 2008).[2] In the United Kingdom, 47,400 people went diving at least once a month in 2008/09 (Sport England, 2009).

Scuba is more expensive to pursue than the NCAs discussed so far in this chapter. The price of a full dive system can be several thousand dollars, while an introductory recreational dive course costs several hundred dollars. Ongoing participation may involve the expense of boat trips to local dive spots as well as more extensive travel to national or international dive locations.

Surfing

The term 'surfing' most commonly refers to the activity of riding waves while standing on a board. In the related activity of bodyboarding, the participant lies on the board with their upper body slightly raised and supported by their arms or uses a 'drop knee' technique, while body-surfers ride waves without the aid of a board. Sea kayakers and coastal rowers may also ride waves, but it is not the exclusive focus of these pursuits and they are discussed separately below, as are windsurfing and kitesurfing which are distinguished by the use of wind power for propulsion.

The 'ancient Polynesian art' of board surfing re-emerged in Hawaii in the early twentieth century and slowly spread around the Pacific Rim, with much of its early growth and development centred on California (Booth, 1996, 2004). In New Zealand and Australia it was, at first, closely associated with the surf life-saving movement which initially arose to oppose Victorian restrictions on bathing in public and then to provide safe conditions for sea bathers in the form of beach patrols and rescue services (Pearson, 1982). This led to the creation of surf bathing clubs and a competitive surfing sport with events to demonstrate a variety of surf-related skills including swimming, boat handling (row boats and canoes) and surf board events, albeit with an emphasis on paddling rather than wave riding (Pearson, 1982). Early wooden boards were heavy and difficult to ride, but with the development in California in the 1950s of the shorter and lighter 'malibu' boards, surfing gained rapidly in popularity. As surfing spread it developed a culture that was increasingly distinct from the discipline and teamwork of the surf life-saving movement, and became more closely associated with the individualism and anti-institutionalism of the 1960s' and 1970s' counter-culture (Booth, 2004; Pearson, 1982). The first International Surfing Championship was held in Hawaii in 1954, and with the advent of competition came another split, this time between those who wanted to pursue competition and those who felt that this was anathema to the true spirit of surfing as 'a communion with nature' (Booth, 1996).

Surfing today is a recreational and competitive NCA involving 'the art of riding a board across the face of a breaking wave' with the ideal wave being 'steep, smooth, high and about to break' (Preston-Whyte, 2002, p. 307). To do this surfers must spot a promising wave and *catch* it by manoeuvring the board, either by paddling or sometimes by *tow-in*, into position in front of it. This requires skill and experience in identifying the wave sets and predicting where they will break. Once the wave has begun to lift the board and push it forward, surfers quickly jump to their feet and begin to *ride* it, endeavouring to stay on the board for as long as possible while *turning*, *carving* and performing other *tricks* depending on their skills. Successive levels of challenge are provided by attempting to catch waves of varying difficulty and under a range of conditions. Bodyboarders differ principally in that they prefer to ride 'hollow, steeper breaks' with their shorter boards (Preston-Whyte, 2002).

The formation of waves is affected by the changeable conditions of the wind and ocean topography. Regular surfers become intimately acquainted with their particular local areas and the variable conditions that impact on the shape and frequency of waves. In his study of surfing in Durban, South Africa, Preston-Whyte (2002) found that local surfers were highly knowledgeable about the nature of the waves found under various conditions in the vicinity and the ways in which they could be surfed. Many surfers are also drawn to travel in search of the 'perfect wave', the image of which is heavily promoted by surfing media and the surfing tourism industry (Ponting, 2008). The marine conditions most likely to manifest a perfect wave are consistent, strong winds over a wide area of open water which 'generate large swell, where the water close to the shore is fairly deep and where the profile of the sea floor rises steadily towards the shore causing the ocean swell to form into steep-sided waves' (Preston-Whyte, 2002, p. 314). In the interests of safety surfers should also be aware of the location of sand bars, rocks, reefs and riptides, as well as the presence of marine life such as sharks, seals, sting rays and jelly fish.

Surfers have an expression for the affective experience of surfing – *stoke*. Surfer Chris Evers (2006, pp. 230–231) describes it as 'a fully embodied feeling of satisfaction, pride and joy. You will tingle from your head to your toes'. In his study of the aesthetics of surfing, Stranger (1999) argues that surfing provides an 'embodied "experience" of the sublime'. He quotes an anonymous surfer:

> When you paddle out and see [a 10 meter high wave] staring you in the face, it's like 'Oh my God'...Being a surfer and being involved with nature all the time gives you a different understanding of where you might find God.
>
> (Stranger, 1999, p. 271)

The International Surfing Association, formed in 1964 and based in California, has a membership spanning 50 countries across six continents. It organizes the World Surfing Games with events in the disciples of surfboard, longboard and bodyboard. Estimates of the global surfing population range from 5 to 23 million (Ponting, 2008). In the United States, around 2.5 million people went surfing in 2009, or roughly 1 per cent of the population, representing an almost 10 per cent increase over the previous decade (Outdoor Foundation, 2010). Australian figures combine surfing and windsurfing to record a total participation rate for surf sports of 2 per cent in 2008, or just over 300,000 people, about 70 per cent of whom participate more than 12 times per year (Australian Sports Commission, 2008). In the United Kingdom, 60,800 people were surfing at least once a month 2008/09 (Sport England, 2009).

A beginner's surfboard can be purchased for a few hundred dollars, while a foam bodyboard can cost as little as 30 US dollars, or a couple of hundred dollars for a more sophisticated model. Additional equipment includes a leash (attaches the board to the surfer's leg), wax (applied to the deck of the board to prevent slipping), a wet suit or rashguard if required, and fins, which bodyboarders may wear for extra propulsion.

Windsurfing

Windsurfing, also known as sailboarding and boardsailing, is a hybrid of surfing and sailing that can be undertaken anywhere there is both water and wind. Its origins have been hotly contested, as prototypes of a board with a pivoting mast and sail attached were developed independently in the United States, the United Kingdom and Australia in the two decades following the Second World War (Reekie, 1996). The design that eventually succeeded commercially was built in California in the late 1960s. Inspired by the search for a cheaper and simpler form of sailing and a faster, less exhausting means of reaching surfable waves, it featured the now familiar universal joint that allows the rig (mast, sail and boom) to be moved in any direction, and the wishbone boom that the windsurfer holds while standing on the board (Reekie, 1996). Foot straps were later added and with shorter boards this made aerial manoeuvres possible, while a chest, waist or seat harnesses can be used to take pressure off the arms. Advances in board design have seen the introduction of a variety of sizes and shapes for different styles and skill levels.

Windsurfing has a reputation for being difficult to master (Wheaton, 2000). Reekie (1996, p. 848) outlines the basic technique, beginning with *uphauling*, whereby a windsurfer balances on the board and slowly pulls the rig off the water and into position using a rope until the boom is reached:

> The hand nearer the front should cross over the other and grasp the boom near the mast and then the back hand grasps the boom about 1 meter (3 feet) back. Pulling in the back hand traps air in the sail and moves the board forward. Steering is performed by raking the rig. Leaning it back pivots the boat more toward the wind, and leaning it forward turns the boat more away from the wind. If either of these maneuvres is carried on long enough, the board will eventually turn completely around – the former is called coming about, and the latter is called jibing. More advanced sailboarders do not uphaul (in fact they cannot, because their boards are likely to be sinkers) but rather waterstart, and neither do they steer with the sail, but rather by banking the board with their feet.

In addition to foot steering, advanced techniques include different turning manoeuvres and various tricks such as jumps, loops and wave riding (Reekie, 1996). Windsurfing offers participants 'a sense of union with the forces of nature', through an exhilarating interaction with waves and wind (Ryan, 2007, p. 105). Dant (1998, p. 83) explains:

> The vertiginous pleasure of windsurfing is derived from the sensations of moving off, accelerating, sailing at speed and making turns, especially while moving fast... Balance is a key motor skill for the sailor who must not only keep her or his own body upright as on a bicycle but also has to counteract the lateral force on the sail

which varies with wind speed. Wind and water surface are continually changing and balance is a continuous process of responding to forces in two planes.

The popularity of windsurfing grew rapidly in Europe in the 1970s and appears to have peaked in the mid-1980s (Reekie, 1996; Ryan, 2007). In the United States, participation in 2009 was down around 35 per cent from 2000 figures, at just over 1 million people (Outdoor Foundation, 2009a). A possible reason for this decline is that its initial appeal was to a young, alternative lifestyle cohort. As this group of windsurfers aged and the sport became increasingly established and commercialized, the next generation of 'alternative youth' have turned to newly emerging activities such as kitesurfing and mountain biking (Ryan, 2007).

Although windsurfing is inexpensive compared with other forms of sailing, it becomes more expensive as you progress with experienced sailors owning multiple boards and sails suited to various wind and water conditions (Wheaton, 2000).

Kiteboarding/surfing

Kiteboarding or kitesurfing combines aspects of surfing, windsurfing and powerkiting (discussed in Chapter 4). The activity had its tentative beginning in the 1970s, when improvements in kite technology led to experiments using kites to pull riders on water skis, windsurfing boards and wakeboards. A breakthrough came in the early 1990s when the American Cory Roeseler developed the kiteski – a single water ski, with a two-lined steerable delta-style kite attached to a control bar and motorized winding system (Beaudonnat, 2006; Boyce, 2004). In 1997 the Legaignoux brothers in France made design improvements by incorporating inflatable tubes into the kite, leading to the commercial production of kiteboards. A year later the first competition was held in Hawaii.

Two main types of kite are used by kiteboarders, Leading Edge Inflatables (LEIs) and foil kites. Boyce (2004) estimates that LEIs account for more than 90 per cent of the kiteboarding market. These are single skinned sails with inflatable tubes – a horizontal one along the leading edge, with vertical battens running down to the trailing edge – which hold the kite's shape and give it buoyancy if it lands in the water, making it relaunchable. Nautical foil kites are based on the parafoil or ram-air wing used by paragliders (see Chapter 2), with the addition of inlet valves to keep the air in the kite when not in flight. This gives it greater stability and means it can be relaunched from the water (Beaudonnat, 2006).

The kite is connected by either two or four lines to a bar held by the kitesurfer and used to control the power and direction of the kite. This in turn hooks into a waist or seat harness, although there are options for disconnecting from the harness to facilitate certain manoeuvres. Most four-line kites have a depowering system incorporated into the control bar that allows the pilot to let out the rear lines, thereby changing the kite's angle of attack such that it loses power. In this way the pilot can adjust the power of the kite for various situations, and it also acts as a safety feature to cope with gust and squalls. As a backup there is usually the additional feature of depowering the kite by letting go of the control bar (Boyce, 2004). Kiteboards were originally mono-directional like surfboards, but later symmetrical designs, or *twin tips*, can be ridden in either direction.

Good conditions for kitesurfing are consistent crosswinds that propel surfers on long, ocean glides (Gluckman, 2008). For smooth flying of the kite, a pilot must understand the impact of land forms on wind speed and direction, and take account of wind effects such as turbulence and whirlwinds caused by obstacles (Beaudonnat, 2006). Kiteboarders need to be able to judge the wind speed, as any kite will have a wind range which indicates the maximum and minimum wind strength in which it can be used. To do this they can check the forecast and observe physical indicators of wind speed such as clouds, flags, trees and water surface. Pocket wind metres can also be used, and clubs or centres will often have an anemometer on site (Boyce, 2004). Tides and currents are other important factors in assessing kitesurfing conditions.

Beginners usually start with learning to control the kite on land, using the control bar to loop the kite in a horizontal figure eight and feeling the kite *power up* as it dives through the centre of their *wind window*. This is something a kite pilot must visualize. Facing down wind with the pilot as a fixed point at its centre, the wind window 'resembles the surface of a quarter sphere.... power kites are most efficient when they are at the centre of the wind window, flying horizontally across the sky at roughly head height or slightly above' (Boyce, 2004, p. 22). As the kite is flown horizontally across in either direction it will lose power as it reaches the edge of the window. Flying up from the centre the kite will also gradually lose power until it reaches the *zenith* above the pilot's head, which is also a safety position that can be used in case of difficulties. Boyce (2004, p. 23) cautions new kiteboarders about the potential of the kites:

Flying a big power kite near the center of the window will generate enormous lateral pull and this is where you'll find yourself leaning right back, even lying down, to stop yourself being pulled over and dragged along on your front. In fact you rarely see...kiteboarders fly their big kites near center window because the lateral pull would be too much to hold, leading to a big horizontal wipe-out....your skill as a flyer will be in learning how to manipulate the kite in the wind window to deliver the kind and quantity of pull you want.

To perform a water start, a kiteboarder must launch the kite and then use the power of the kite to get into a standing position on the board. To ride across the water the pilot stands in the centre of the board and is pulled in the direction that the kite is flown. To increase speed a pilot flies figure eights, with the kite building up the most pressure as it goes upwards and downwards through the wind window. The stance on the board is also used to control direction and speed. With twin tip boards it is not necessary to jibe, making changing direction more straightforward. Other necessary skills include slowing down and stopping by piloting the kite into the correct position, leaning back and applying heel pressure. Advanced skills include jumping and other aerial manoeuvres, rotations, handle passes and wave riding.

The world speed record on a kiteboard is just over 50 knots (or more than 90 km/h), which is slightly faster than the windsurf record and faster than all other sailcraft apart from the hydroptère (http://www.sailspeedrecords.com/500-metre-records.html, retrieved 1 June 2010). Kiteboarders can also become airborne for seconds at a time and many participants prefer it to windsurfing (Gluckman, 2008). As one convert puts it, 'I switched to kiteboarding (from windsurfing) for the big airtime. Combined with less hassle, kiteboarding has the best bits of windsurfing times ten' (Boyce, 2004, p. 108). Boese & Spreckels (2008, p. 7) describe the attraction:

> ...kitesurfing is the sport that combines the elements. The boundary between water and air dissolves in a fluid transition: for seconds at a time a kitesurfer is aloft, floating above the water on his kite and experiencing a sense of virtual 'zero gravity' before landing on the ocean again, only to take off once more after brief contact with the water. Or he rides the wave on his small board, so that time seems to stand still. A feeling of being at one with oneself and with nature is produced; boundaries melt away. It is this feeling that drives kitesurfers...they love the sea, the sky so close above them, and the

wind, which bestows power, propulsion and perhaps even something a little regal.

The main kiteboarding competition disciplines are: freestyle (tricks, aerial manoeuvres, rotations, handle passes and so on); hang time (length of time in the air); wave riding (similar to surfing, including *tube riding*); and kitespeed (Boyce, 2004). The global governing body for the sport is the International Kiteboarding Association, which is an International Class Association of the International Sailing Federation (ISAF).

Participation in kiteboarding is relatively low but said to be on the rise (Gluckman, 2008). In France there were around 7000 kiters (including land, snow and water) in 2009, with steady growth over the past 5 years (http://federation.ffvl.fr/sites/ffvl.fr/files/Plaquette%20statistiques%202009%20-%20def.pdf, retrieved 10 April 2010). In the United Kingdom in 2008/09 11,600 people were kiteboarding on a monthly basis (Sport England, 2009).

An inflatable kite and quick release harness can cost over a thousand dollars. As with windsurfers, kiteboarding devotees may have a number of different sized kites, each suited to particular wind strengths (Gluckman, 2008).

Sailing

Sailing, as a form of transport, dates back thousands of years, but its history as a sporting and recreational pursuit is traced to The Netherlands in the seventeenth century, from where it was introduced to England. The first sailing clubs were formed in the early eighteenth century and the initial focus was on racing, with participation being largely the preserve of the wealthy (Jennings, 2007b). With the development of smaller boats such as the dinghy in the late nineteenth century and cheaper fibreglass, mass-produced yachts in the mid-twentieth century, sailing gradually became accessible to a much wider range of people and was promoted as both a competitive and recreational activity (Aversa, 1986). Today sailors can be found recreating wherever there is a large enough body of water: out at sea, in harbours, on lakes and even large rivers. These recreationists can be classified, according to different levels of participation, as *day sailors*, *short-term cruisers* and *cruisers*, or those on long-term and sometimes open-ended voyages around the world (Macbeth, 1985 in Jennings, 2007b).

All sail boats have a rig, comprised of at least one mast and one sail. Their hulls come in various hydrodynamic designs, incorporating some form of underwater foil such as a keel or centreboard to prevent them from slipping sideways or capsizing. A dinghy is a small, open boat and most are about 4–5 m long. A catamaran is similar to a dinghy but has two hulls. Keelboats are open like dinghies, but larger (around 6–9 m) with a keel. Yachts are keelboats with cabins and accommodation and may be designed for coastal and ocean cruising, or racing (Evans et al., 2007).

Using the aerodynamic force of the wind on the sails, and the hydro-dynamic force of the underwater section of the hull, a sail boat can move forward in almost any direction except within about 45 degrees of upwind. A sailor manages the force of the wind in the sails by alter-ing the angle of the boat and the angle of the sails relative to the wind. A well *trimmed* sail will be neither too loose – causing flapping – nor too tight, which will disrupt the airflow over the sail and slow the boat (Evans et al., 2007). This requires continual adjustment in response to shifts in the strength and direction of the wind. To go upwind a sail boat must *tack*, turning the bow of the boat through the eye of the wind to change direction. Jibing (or gybing) is the same manoeuvre when sailing downwind, with the stern turning through the eye of the wind.

Being light and small, dinghies are particularly responsive to wind and water conditions and thus provide 'an intimate sailing experience' (Evans et al., 2007, p. 40). This makes them ideal for learning the basics of sailing and experiencing its pleasures. As they have a move-able centreboard rather than a weighty fixed keel, dinghies are quick to start *planing*, whereby 'the hull lifts on to its bow wave and leaves its stern wave behind. As the boat accelerates on to the plane, its bow lifts clear of the water, its wake becomes flat, and it rides on the flat-tish aft part of its hull' (Evans et al., 2007, p. 48). Dinghy sailors use their body weight to *trim* the boat and keep it upright against the force of the wind and the water conditions. This may involve mov-ing their weight over the side by *hiking* with their feet hooked under toestraps or using a *trapeze* which allows them to stand on the side of the boat with their bodies suspended above the water. With expe-rience, sailors know intuitively how to weight the boat under various conditions. Irini Kotroni recounts the joy of day sailing a dinghy in the Mediterranean:

> Beneath us is the reef; the sea changes colour from blue to topaz with sparkling sunlight dancing on the water. We head down the path of the sun. It is extremely hypnotic; surely we could sail until the sun goes down? Enticing it may be, but we have to remember that if the

wind dies, we paddle back. We jibe or tack along the coast, gauging the best areas for the wind to catch our sails.

As the wind increases and the sailboat begins to heel over, I hook the trapeze to my harness, pushing myself out. I lean back at full stretch to bring her upright. We accelerate and suddenly we are 'harnessing' the wind, skimming over the surface of the water as the boat lifts and planes. It is as if we are riding a thoroughbred racehorse down the final furlong to win the race.

The wind and spray blow through my hair, thankfully my sunglasses and cap are firmly strapped to my head. My whole body is exhilarated; I want to 'shout' and 'whoop' with joy at the sheer excitement of it all.

(www.helium.com/items/1791965-the-joy-of-sailing,
retrieved 20 May 2010)

With their heavy keels, most yachts are not able to plane and the crew's weight usually has little impact on their performance, except on some racing boats. Typically they have a form of auxiliary power, used for docking and for charging the batteries that power onboard electrical systems. While yachts are sometimes sailed single-handed, more often they are managed by a small crew, particularly under the demands of coastal and ocean cruising when they face the challenge of rougher conditions. Beyond basic sailing skills, cruising requires extensive knowledge of meteorology and navigation at sea, marine radio communications, off-shore safety and sea survival. Jennings (2007b) has found that those who have chosen a cruising lifestyle are motivated by a desire to experience natural environments *in situ* and being able to explore out-of-the-way places, as well as the fun and challenge of sailing.

Training in various aspects of sailing, including navigation and seamanship, is offered by various clubs and associations around the world, as well as certifications such as the Royal Yachting Association's Yachtmaster Certificates of Competence. The International Sailing Federation, established in 1907, governs the sport globally and sets the rules and regulations for sailing competitions, and runs world championship events.

Aversa (1986) finds evidence of increased participation in sailing in the United States between 1960 and 1980 from boat sales figures and a growth in clubs. However, Kelly and Warnick (1999) report a 'long-term but irregular decline' for US participation rates since 1979, with around 5 million sailors in the mid-1990s, 20–25 per cent of whom were frequent participants.[3] In 2009 there were 4.3 million sailors in the United States (around 1.5 per cent of the population), representing a 1.4 per cent decline since 2000. In Australia 112,000 people went sailing and/or windsurfing in 2008 (Australian Sports Commission, 2008), and in the United Kingdom 175,000 people were sailing monthly in 2008/09 (Sport England, 2009). Declines in sailing participation are possibly linked to its relatively high costs and the rise of cheaper alternatives such as windsurfing in the 1980s, and more recently kitesurfing, replacing day sailing as an activity. Cruising requires a significant investment in the purchase of a boat (or the hire of one for shorter trips), as well as having the time and financial resources to sustain a long voyage.

Canoeing and kayaking

Early kayaks and canoes were made from natural materials such as wood, bark and animal hides, and have been used for hunting, fishing and transportation over thousands of years on the rivers and oceans of the Americas and Polynesia (Harrison, 1993). Recreational canoeing emerged in the mid-nineteenth century when it was used for wilderness trips and offered as an activity at resorts (Townes, 1996a). Its popularity in the United Kingdom and Europe was boosted by the Scotsman John MacGregor, who wrote books about his paddling adventures around the world and formed the Canoe Club in the 1860s (Hudson & Beedie, 2007; Townes, 1996a). Since that time, design innovations and new materials have significantly improved the portability, manoeuvrability and durability of canoes and kayaks. In the late 1960s very affordable moulded plastic boats became available and these now dominate the market (Hudson & Beedie, 2007; Townes, 1996a). Today highly specialized designs cater for the many variations of this NCA that have evolved.

Canadian canoes are generally open, although they may be decked, particularly for use in whitewater, and are manoeuvred with a single-bladed paddle, sometimes from a kneeling position. Kayaks are a type of canoe that originated among the Inuit cultures of the Arctic. They typically have an enclosed deck and are paddled in a sitting position with a double-bladed paddle. A *spray skirt* seals closed the cockpit around the waist of a kayaker. Kayaks are narrower than canoes and their centre of gravity is lower, giving them the same stability as a canoe while enabling greater speed.

One of the charms of canoes and kayaks is that they are elegant and simple, and yet, it would seem, endlessly versatile (Kuhne, 1998). They can negotiate a wider range of water than most other boats – anything from 'whitewater rivers, big, slow-moving rivers, tidal estuaries, coastal bays, shorelines, inlets, lakes, inland seas (like the Great Lakes), and the open ocean' (Harrison, 1993, p. 2). As Kuhne (1998, p. 1) puts it, a canoe:

> ... is sleek enough to glide easily through water, yet capable of haul-ing heavy loads; it is manoeuvrable in whitewater, yet able to hold a straight line in the wind; it is rugged enough to withstand abuse, yet light enough to be portaged.

The different types of canoeing and kayaking practised today include: sea kayaking, whitewater river descents, multi-day camping trips (in coastal, lake and river settings), competitive events (slalom, marathon and river racing), kayak surfing and play-boating. Canoeists and kayak-ers feel the dynamics of the water through their boats as if it were touching their skin and become adept at responding to these chang-ing dynamics to keep the kayak upright (Mattos & Middleton, 2004). Various paddling strokes are used to propel the canoe forward or back-ward, to turn it and brace for extra stability. Skilled paddlers can execute these strokes with smoothness and efficiency, using the strong muscles in their torsos, as well as their arms and shoulders (Kuhne, 1998).

Whitewater, or wild water as it is sometimes called, refers to the tur-bulent sections or rapids usually found in steep rivers, where the water flows quickly over an uneven river bed, causing a chaos of waves and *holes* (Mattos & Middleton, 2004). The availability of shorter plastic boats led to the growing popularity of *running* whitewater in the 1970s (Townes, 1996a). Enthusiasts learn to *read* whitewater; that is, to assess a stretch of river in order to determine its various obstacles and hydraulic features, and thereby the best way to tackle it (Kuhne, 1998, p. 68). This may require *scouting* the rapids – walking downstream on the riverbank – to get a better view of what lies ahead, and trying to identify a potential route through it. Making such an assessment requires an understanding of the way in which the volume, gradient and form of a river creates par-ticular kinds of waves, holes (hydraulics strong enough to stop and hold a boat) and currents. Whitewater kayakers must learn how to launch their boats and enter the rapids (*breaking in*), to ferry glide across the river and avoid obstacles, and to *punch* through difficult hydraulics if they are unavoidable, as well as how to paddle over various kinds of vertical drops. The calmer waters downstream of protrusions, known as eddies, can provide a place to rest and reassess the route below. The key to staying in control in whitewater, according to Mattos & Middleton (2004, p. 31), is anticipation: 'If you react to the water, your actions will always be too late. You must lean, edge and use the paddle, anticipating what the water will do. This comes only with experience.' The need for intense concentration and to perform spontaneously and automatically has, not surprisingly, been associated with the experience of a state of flow in whitewater kayakers (C. D. Jones et al., 2000). The special skill, knowledge and experience required in whitewater kayaking has also been observed by Stebbins (2005a) in the only study of this NCA undertaken from the serious leisure perspective.

Knowledge of rescue techniques is also essential in whitewater. If a kayaker capsizes, they will attempt to roll back over using one of a number of methods involving body and/or paddle movements. If a roll is not possible the paddler must exit the boat and either swim or be towed to safety by another boat, or be pulled to the river bank by a rescue rope or *throw-line*. If it is decided not to paddle certain rapids, the boat can be *portaged* (i.e., carried) around the section of river, or guided along the bank attached to a rope.

In play-boating, the whitewater paddler's skills are exercised for the pure pleasure of 'frolicking in the waves and hydraulics, surfing, and performing acrobatic tricks' that exploit the power of the water (Mattos & Middleton, 2004). Squirt boating, waveski and kayak surfing are other playful variants of canoeing.

In open water and sea paddling the emphasis is on travelling over long distances and exploring bodies of water and their adjacent environments. While kayaks are particularly well suited to coastal and ocean paddling, open canoes are typically used for journeys on large rivers and lakes. The requirements for long distance paddling include speed, directional stability and self-sufficiency (Mattos & Middleton, 2004). Open water kayakers or canoeists must be able to negotiate tides and currents, wind and waves and be skilled at navigation. The strokes of an open water paddler require 'more finesse and subtler leans' than their whitewater counterparts – rather than aggressive anticipation, sensitive responsiveness is required (Mattos & Middleton, 2004). Ideal locations for this variant include stretches of open water 'interspersed at short intervals with interesting shapes of land – islands, inlets, coves, bays, cliffs, beaches, and other places to land the kayak' (Hudson & Beedie, 2007, pp. 178–179). The experience of the surrounding natural environment and its wild inhabitants is of particular importance to long-distance paddlers and the 'corporeal' challenge of handling the boat in the water mediates the landscape which is traversed and shapes its significance (Mullins, 2009). The resulting sense of place is vividly captured in this reflection by canoeist Cecil Kuhne (1998, p. 130):

> A journey by canoe is a passage from one life to another. You leave the noisy and predictable routine of the metropolis for the quietude and uncertainty of nature. The deeper you paddle into this world, the more your perspective changes. You see new things, of course, but you hear, smell, and feel them with an intensity you never thought possible.
>
> ... I vividly recall an alpine lake whose banks seemed to heave straight towards the heavens. Sunshine streamed into the narrow gorge to warm the white beaches spotted with lichen-stained rocks. Side canyons, filled with ferns and berry bushes, occasionally pierced the chasm to allow creeks to flow through. Along the shore lived deer, elk, bighorn sheep, and mountain goats. High above flew ospreys and bald eagles.
>
> I remember paddling the wind-swept ridges and crags of a desert stream. There the gentle flow of the river piled itself against precipitous cliffs, where light danced upon the swirling surface. The air was as dry and brittle as the canyon that etched the cobalt sky. You could feel on your face the gentle, unceasing wind. The silence there was profound. At night we gazed up at that narrow swath in a pitch black sky with stars close enough to touch – and the rest of the world seemed very far away.

The International Canoe Federation, founded in 1924, has over 150 member countries. Figures from the United Kingdom and Australia show similar participation rates of around 0.35 per cent of the population canoeing and kayaking an average of once a month, and an increase of almost 40 per cent in the Untied Kingdom over the past few years (Australian Sports Commission, 2008; Sport England, 2009). In the United States in 2008 almost 10 million people went canoeing and another 9 million went kayaking, making it one of the most popular NCAs in that country, with more than 5 per cent of the population participating. While kayaking has fewer participants than canoeing, numbers have grown steadily since 2006, particularly for whitewater kayaking, and kayakers paddle more frequently than canoeists. About two-thirds of kayakers described themselves as recreational, with the remainder being whitewater or sea/touring participants (Outdoor Foundation, 2009b).

Canoes and kayaks are relatively cheap and transportable watercraft. A 'starter package' for a beginner ranges from around $1000 for surf kayaking to $2000 to $3000 for white water equipment (including boat, paddle, spray skirt, paddle jacket, flotation device and air bags). Skills courses are provided by clubs and commercial kayaking centres.

Whitewater rafting

Rafting is another traditional form of water transport which has become popular for recreation. A raft is essentially a 'floating platform' which may be manoeuvred by oars, paddles or an outboard motor (Addison, 2000). In comparison to other types of boats they are stable, flexible and ideal for shallow, rough water due to their capacity to withstand currents and hydraulics, and rebound off rocks. They are also capable of carrying heavy loads of people and/or equipment. On the other hand, they create more resistance in the water and are more difficult to steer than canoes or row boats (Townes, 1996c). Rafts can be used to navigate flat water and make ideal transport for anglers, hunters and nature photographers (Townes, 1996c). However, whitewater river descents in inflatable rafts, which have become popular since the 1960s, are the main way in which rafting is engaged in as an NCA.

The most common types of inflatable raft used for whitewater rafting are the oar raft, the paddle raft, the cataraft and the inflatable canoe. Oar rafts have a frame, attached to the top of the tubes, to hold the oars. They are typically rowed facing downstream using a special pushing stroke called the *portegee* position to manoeuvre through rapids. In a paddle raft everyone onboard paddles in a coordinated fashion, while a skipper sits at the stern steering the raft and calling out instructions to the crew. A cataraft has two inflatable tubes connected by a frame and may be set up to be rowed or paddled. A small cataraft can be controlled by a single person with oars. One- or two-seater inflatable canoes, which can be used with double-bladed paddles, are suitable for smaller rivers and very stable and manoeuvrable.

The key rafting manoeuvres are tracking (straight forwards or backwards), turning, pivoting (rotating on one spot), ferrying and cresting (over the top of waves) (Addison, 2000). Another key technique is *highsiding* which is used to stop the raft from flipping over if it hits a wave or a rock or becomes stuck in a hole. This involves the whole crew throwing their weight towards the side of the raft that is being lifted by the obstacle. Strong strokes can then be used to pull the raft over the rock or wave, or out of the hole.

According to Addison (2000, p. 61), whitewater rafting requires a 'panoramic vision' that takes into account 'the boat and crew, the micro-currents, the macro-currents, and the surrounding environment including the lie of the land and the weather'. Having assimilated all this information, a rafter must then determine exactly the right move to maintain the desired line through the rapids and execute it with precise timing and control: 'A deft oar stroke or planting your paddle in an eddy beside the boat can ensure a faultless run, although the water all around may seem chaotic' (Addison, 2000, p. 40). Patricia McCairen (1998, pp. 170–171) reflects on what she has learnt from rafting and how this helped her negotiate the infamous Crystal Hole during a solo journey through the Grand Canyon:

> As a beginner, I'd sit in an eddy behind a boulder in the middle of a rapid, fascinated by the water's motion...As my rowing improved, I came to understand that using the river's strength, direction and flow helped me more than absolute control or total passivity... [I] walk along the edge of the river, studying the rapid, picking out a particular curl of a wave, a pillow of water as is flows up on a rock. I examine each subtlety until it is established firmly in my memory, and there is no chance I will get lost when I begin my run.
>
> Back at *Sunshine Lady*, I tuck things away, untie and gently shove off. I float slowly downstream, holding the raft just outside the eddy that hugs the right shoreline. As I near the head of the rapid, I pivot the raft, turning the stern at a forty-five degree angle to the current, which gives me additional power when I pull the oars. It is at this moment, on the edge of the rapid, that everything else disappears. I am totally focused – nothing else matters, nothing else exists. Everything blends together: Even *Sunshine Lady* no longer exists of her own entirety, but instead becomes part of me, and the two of us together become part of the rapid.
>
> I hug the right shoreline, skirting smooth stones and small holes, neither pushing the river nor allowing her to sweep me away. I pass Crystal Hole with more than a boat-length to spare, awed by her power and majesty.

Whitewater rafting trips can be day trips or multi-day expeditions. On longer trips, as with canoeing and kayaking, the challenge of negotiating rapids is interspersed with sections of free-flowing water on which the remoteness and isolation of canyons, waterfalls and other scenic aspects of the river setting can be enjoyed (L. Jones, 2007). Travelling through canyons can be a particularly awe-inspiring experience due to their vastness and the sense of solitude and isolation they evoke. In their ethnographic study of commercial rafting trips in the Colorado River basin, Arnould & Price (1993) found that 'communion with nature', including aspects of the water, geology and wildlife, was a significant aspect of the trip, while the teamwork required fostered a strong sense of 'communitas'.

Most organized rafting is centred around team competition and commercial guiding operations. Some canoe, kayak and general whitewater clubs include rafting among the paddling disciplines their members participate in, such as the Kansas Canoe and Kayak Association whose members organize rafting trips. The International Rafting Federation was formed in 1997 in response to a growing interest in competitive rafting and the need for an internationally recognized guide training and award scheme. The Rafting World Championships and other regional competitions have been held since the late 1990s with national teams competing in sprint, slalom, head to head and down river racing events (see www.internationalrafting.com).

In the United States rafting has lower participation levels than canoeing and kayaking, with 4.7 million rafters in 2008. Frequency of participation is also lower, with 43 per cent of rafters getting on the water only once a year and only 8 per cent rafting more than 12 times annually (Outdoor Foundation, 2009b). Monthly participation in the United Kingdom is also low at 5700 in 2008/09 (Sport England, 2009). These figures suggest that rafting attracts predominantly casual participants, many of whom probably take part in commercial trips. For example, more than 21,000 people raft through the Grand Canyon each year, but only 15 per cent of these are private rafters (L. Jones, 2007). However, demand for private permits to raft the river is growing. In 2003 there were more than 8000 people on the waiting list and the Grand Canyon Private Boaters Association (www.gcpba.org) has been formed to advocate for improved access for non-commercial rafters.

A raft can cost between 2000 and 6000 US dollars depending on the model, although smaller catarafts and inflatable canoes can be acquired for less. The expense of purchasing a raft, and the inconvenience of transporting it to desirable location, may be one reason why many rafters prefer to hire equipment from commercial outfitters or pay to join organized trips.

Rowing and sculling

Throughout history, rowing has been important for the purposes of transport, trade, fishing and warfare. Its origins as a modern competitive sport and NCA are traced to eighteenth-century England, when races were organized between professional 'watermen' on the River Thames. By the early 1800s, boat races were being held in Australia, Russia, Canada and elsewhere in Europe (Roberts, 2007). As the sport evolved in the nineteenth century, boats were developed to make them faster and more efficient to row: outriggers allowed the oars to be mounted in oarlocks outside of the gunwales, which meant that hulls could be narrower and lighter while still permitting the same leverage in a stroke; and a sliding seat permitted the rower's leg power to be incorporated into the stroke (Redgrave, 1995, p. 4). By the end of the nineteenth century, clubs were being formed throughout Europe, North and South America, Australasia, South Africa, Russia and Japan (Roberts, 2007).

A rower sits in a row boat facing backwards and propels the boat forwards using oars held in place by oarlocks. In *sweep rowing*, each rower has one oar held with both hands, while *scullers* hold two oars, one in each hand. Sweep rowers may row in pairs, fours or eights, while scullers row in singles, doubles and quads. All eights also carry a coxswain, who steers the boat and coaches the crew while on the water. Coxswains are optional for other sized crews and are less common in sculls. In coxless sweep boats, one rower will steer using a foot-rudder, while sculls are not equipped with a rudder and steering is achieved by increasing or decreasing the power applied to one side.

The various types of rowing boat include the traditional wherries which, being relatively heavy and originally made of wood, are wider and shorter than later designs. Open-water/recreational boats are designed for safety, stability and seaworthiness, with high gunwales and built-in flotation. Racing shells are 'slender, light, responsive, high-performance boats used almost exclusively on protected lakes and rivers... [and] tougher to row in rough water' (Paduda, 1992, p. 20).

The rowing stroke consists of four phases: the *catch* (when the oars enter the water), *drive* (pushing with the legs against the foot block and pulling the oars towards the chest), *finish* (the point at which the oar is extracted from the water) and *recovery* (sliding forward to bring the body and oars back into position for the catch) (Redgrave, 1995). As Paduda (1992, p. 13) explains, the stroke should be performed as a 'continuous movement, with no part of the cycle rushed or jerky'. This challenge makes rowing 'an intense sensory and analytical experience':

You feel the wheels of the slides turning under you, pressure on the tops of your feet as you pull yourself up to the catch, your weight hanging on the oars as the blades bite the water the instant the drive begins. You hear your labored breathing, the water murmuring under the boat, and the soft sounds of the catch and release that announce your efficiency. You sense the wind pressing on your blades as you roll them up to the catch, and see the water ripple.

(Paduda, 1992, p. ix)

Rowers must execute their stroke while maintaining the stability of the boat in the water, with the light, narrow, hydrodynamic racing boats having the least inherent stability. When rowing as part of a crew, there is the additional challenge of co-ordinating the stroke rhythm. Indeed, with the exception of single sculls, teamwork is a central element of rowing, with much of the pleasure coming from working in unison (*in swing*) with others to achieve the single goal of 'moving along the water in an effortless glide' (Roberts, 2007, p. 6).

Most rowing takes place on relatively flat bodies of water such as lakes and reservoirs, or rivers with gentle currents. Coastal and ocean rowing are more demanding variants, where rowers face additional challenges, as world coastal rowing champion Jean-François Malval explains:

Knowing how to adapt to weather conditions and row with the best possible technique when the winds are contrary and the waves are swelling, when there are side and tail winds, and when there are currents... We also need to anticipate what could happen and study the sea chart before going out on the water to be able to make the most of the water surface we have. Knowing the tide and wind direction is essential.

(http://www.worldrowingmagazine.com/
worldrowingmagazine/200803/?folio=15#pg17,
retrieved 9 May 2010)

The Fédération Internationale des Sociétés d'Aviron (FISA) was founded in 1892 and now has a membership of 128 National Rowing Federations. In the United States there are more than 1000 registered organizations representing around 80,000 competitive rowers (Roberts, 2007). In recent decades, the focus of most clubs has been competitive activities but recreational rowing is experiencing a resurgence, with the likes of the Amateur Rowing Association in the United Kingdom starting to promote non-competitive rowing, including long-distance and touring rowing (http://www.britishrowing.org/recreational, retrieved 2 May 2010). Since 1971 FISA has been organizing annual World Rowing Tours in different countries. Suisse rower Hans Sprunger describes a tour on the Nemunas river in Lithuania in 2009:

> For me personally it was a wonderful experience – a river that is free flowing (for most of the part), calm waters, green shores and silence.
> There were no motorboats, no planes above, no roads or railways running along the river. Seeing water birds, jumping fish or quiet fishermen and getting into the rhythm of rowing with friends is an experience that will not be forgotten so quickly.
>
> (http://www.worldrowing.com/display/modules/news/
> dspNews.php?newid=324728&pageid=49,
> retrieved 1 May 2010)

Coastal rowing also appears to be a growing aspect of the sport with the Irish Coastal Rowing Federation, for example, reporting a 10 per cent annual increase in membership (http://www.worldrowing magazine.com/worldrowingmagazine/200803/?folio=15#pg18, retrieved 9 May 2010). Total participation in Australia in 2008 was 53,800 (Australian Sports Commission, 2008), while monthly participation in the United Kingdom was 99,100 in 2008/09 (Sport England, 2009).

A boat is the most expensive piece of equipment for rowers, with a single scull boat costing several thousand dollars, and eight-person 'shells' running into tens of thousands of dollars. Purchasing second hand and using club boats is a means of keeping the cost down for individuals, and once access to a boat is ensured there are few additional costs.

Power boating

Power or motor boats are vessels with a motor as the primary or sole means of propulsion (Jennings, 2007a). They range from small runabouts with modestly powered outboard motors through to luxury yachts with high performance jet engines. The history of motorized boating is intertwined with the evolution of recreational boating and the development of engine technology in the late nineteenth and early twentieth century. While early versions used steam and electric-driven engines, it was the internal-combustion engine that provided the breakthrough for recreational and competitive motor boating, which both grew rapidly in popularity in Europe and North America after 1900

(Choate, 1957; Jennings, 2007a). The decade following the Second World War and again after 1970 were further growth periods for boating, although there are indications of recent declines in boat ownership (Richins, 2007; Townes, 1996b).

Power boats may have monohulls or twin hulls, while the shape of the hull affects the way the boat handles in the water. Design variations are a trade-off between speed and stability, with an emphasis on either *displacement* or *planing*. Boats with deeper and broader hulls which displace the water as they move through it are slower, but have greater stability and seaworthiness. Those with shallower, planing hulls designed for *dynamic lift* are capable of high speeds, but are less stable and potentially more difficult to control (Townes, 1996b). The motor may drive one or more propellers, or it may draw water from underneath the boat which is then expelled from a nozzle at the rear of the boat, as with jetboats. These highly manoeuvrable watercraft were developed in New Zealand in the 1950s as a means of navigating shallow rivers which were unsuitable for conventional propeller driven boats.

The key to handling a motor boat is to appreciate that the power and steering control are in the rear or stern of the boat, so in making a turn, the stern swings to one side. In addition, the boat will slip sideways when turning, even if it has a full keel, and a boat with a single propeller will experience a degree of side thrust as well as forward thrust caused by the rotation of the propeller. To stop a boat you can either cut the power and drift to a stop or apply reverse thrust. These characteristics can make manoeuvring in small areas, when docking for example, particularly challenging. Motor boats are also affected by wind and current – in particular by wind as they generally have more bulk above the waterline than below. On the open water a major concern is adjusting the boat's *dynamic trim* (its attitude in the water) to the conditions, as this affects planing efficiency, steering, comfort and safety (Sweet, 2006). Mastering these various facets of boat handling is the core challenge of power boating, as one enthusiast explains:

> Learning exactly how your boat behaves in every situation is – or ought to be – a never-ending process. The learning is rarely a chore, and it's deeply satisfying to know intimately a particular boat's every nuance of behavior.
>
> (Armstrong, 2005, p. 104)

The attractions for power boaters are 'the enjoyment of being out on the open water and the challenge of manoeuvring their vessel on its constantly changing surface...[and] the excitement of operating powerful motorized vehicles at high speed' (Townes, 1996b, p. 646). This NCA is often undertaken as a group activity with family and friends. It provides access to wilderness locations and can be combined with and facilitates other water-based activities. Research by the Recreational Boating and Fishing Foundation in the United States found that 42 per cent of boaters go boating in order to go cruising and about a third to go fishing. A further 13 per cent participate in other activities such as water skiing, tubing and kneeboarding. The main reasons people gave for their participation included release from stress, a restful experience, and the opportunity to connect with nature. Boaters also associate the experience with feeling excited and invigorated, and feeling the warm sun on their face and the wind in their hair (http://www.rbff.org/uploads/Research_section/a-partII.pdf, retrieved 23 May 2010).

A personal watercraft (PWC), often known by the brand-name jet ski, is a monohulled boat designed to be operated by one person standing, kneeling or sitting. It is steered from the front, and has an inboard engine powering a water-jet pump. While its advantages include 'affordability, ease of use, handling and manoeuvrability, exhilaration and excitement', sales of PWC in the United States reached a plateau in the mid-1990s and have been steadily declining since (Richins, 2007).

The Union Internationale Motonautique (UIM) is the international governing body for power boating. Formed in 1927, the UIM now has 52 members. Competitive power boating includes races involving two or more contestants and individual attempts on speed records.

In the United States in 2000 it was estimated that there were between 22 and 24 million adult boaters, and boat registrations grew by almost 14 per cent during the 1990s (http://www.rbff.org/uploads/Research_section/Lit-Review-Final.pdf, retrieved 29 May 2010). However, there is concern that boating participation is not keeping pace with population growth and that changing demographics will lead to further declines (http://www.rbff.org/page.cfm?pageID=8, retrieved 29 May 2010). The 2002 National Recreational Boating Survey found that the most popular type of recreational power boat in the United States is the open motor boat, with over half of boaters using this type of craft, followed by PWC[4] (24 per cent) and cabin motor boats (11 per cent) (http://www.rbff.org/uploads/Research_section/USCG_NRBS_2002-Report.pdf, retrieved 28 May 2010). In Australian in 2008, 180,500 people went power boating, jet skiing and/or water skiing, and about 40 per cent of those participated more than 12 times in the previous year (Australian Sports Commission, 2008).

Power boaters have to meet not only the initial costs of acquiring a boat, but also the ongoing costs of fuel, maintenance and repair. Depending on the size of the boat and access to the water there can be additional expenses incurred for storage and transportation (Jennings, 2007a).

Water skiing and wakeboarding

The advent of power boating has facilitated the development of towed water sports, beginning with water skiing and expanding into numerous variations including barefooting, wakeboarding, wake surfing and skating, kneeboarding and hydrofoiling. In this section we focus on water skiing and wakeboarding as the main towed water-based NCAs.

Water skiing first appeared in the 1920s, and Ralph Samuelson of Minnesota in the United States is widely credited with having developed the idea of being towed behind a motor boat on two skis. The challenge for a water skier, as Favret & Benzel (1997, p. 65) explain, is 'to perceive, interpret, and perform a variety of movement combinations with accuracy, finesse, timing, and power.... awareness and perception [are] required to interpret environmental factors such as wind and water conditions, to assess the performance of equipment, and to determine the skill movement best suited for the situation'.

Starting in the water in a crouching position and holding the ski rope with the tips of the skis pointing up, skiers must hold their position and stay balanced while the boat accelerates and pulls them out of the water. They must then maintain their balance while turning and in response to changing boat speeds or water conditions, by making either major, or very subtle, body adjustments. Water skiers use *edge control* to manage the interaction between the skis and the water's surface, the force of which would otherwise cause the skis to turn or rotate. Two methods of control are used: *inclination*, which involves a taller stance and is used mostly by slalom skiers, and *angulation*, which utilizes the hips, knees and ankles to create angles and set the edge, and is used by most jumpers and trick skiers (Favret & Benzel, 1997). Skiers also use *pressure control* to adjust the pressure of the skis on the water as they move through it. This works with edge control to maintain the angle and direction of the ski – the greater the angle, the greater the pressure. When turning pressure must be smoothly shifted to the outside ski or edge, with advanced skiers making more aggressive weight transfers.

Slalom skiers use body rotation to make turns on one ski along a slalom course, with the resulting accelerations and sharp turns providing an exhilarating experience. Trick skiing is done at slower speeds and requires awareness of body position and precise execution of movements to perform flips, twists and turns, and jumps are made using ramps.

Wakeboarding evolved in the 1980s from combining aspects of surfing, water skiing and snowboarding. Slower speeds are required to start planing on a wakeboard and more resistance is met when making a water start. Wakeboarders perform jumps and other aerial manoeuvres by hitting the boat's wake, as well as using sliders (stationary structures riders slide along) and kickers (similar in concept to water ski jumps) (www.wake.co.nz/features/article_whatis.php, retrieved 29 May 2010).

The three traditional events of competitive towed water sports are tricks, slalom and jump. Participation figures from the United States show 5.6 million water skiers and 3.5 million wakeboarders in 2008, totalling more than 3 per cent of the population (Outdoor Foundation, 2009a). Monthly participation in water skiing in the United Kingdom in 2008/09 was 14,400 (Sport England, 2009).

Beyond the purchase of the basic equipment, a major component of the cost of towed water sports is associated with the boat. A typical wakeboarding boat, for example, costs around 40,000–50,000 US dollars. On top of that are the ongoing costs of boat ownership mentioned in the previous section. A cheaper, simpler and perhaps more environmentally sustainable alternative is cable water skiing and wakeboarding where participants are pulled by an overhead cable system constructed over a body of water. There are a growing number of cable parks around the world, which charge customers hourly or daily rates as well as offering monthly and yearly passes (http://cablewakeboard.com/pages/what-is/, retrieved 24 May 2010).

Conclusions

The relationship between water-based NCAs and the natural environment is multi-faceted. While a range of environmental impacts have been identified across the spectrum of activities, many participants are active in not only trying to mitigate these impacts, but also in pursuing broader agendas to promote water quality and the integrity of the aquatic environment upon which their enjoyment depends.

The negative impacts these NCAs can cause include disruption and dislocation of wildlife, erosion and pollution. Nesting birds, sea mammals and spawning fish in particular are sensitive to the noise and proximity of motors and people, which can lead to disturbed breeding patterns and dislocation. Turbidity caused by propellers and jet engines stirring up bottom sediments in shallow water, or from divers disturbing the seabed, can cause further disruption. Seabed damage, especially on fragile coral reefs, may also result from anchoring, groundings and the impacts of snorkelling and scuba activities (Jennings, 2007b). Erosion of shoreline vegetation can be caused by boat wakes and camping activities on multi-day canoeing and rafting trips. Polluting agents may enter the water via fuel leaks or spills from power boats and sail boat motors, as well as the use of non-biological cleaning agents and the careless disposal of sewerage and other waste (Jennings, 2007b). The emission of hydrocarbons from motor boats and biological contamination from transporting invasive organisms between locations on equipment are further impacts. Another challenge is the demand for jetties and marinas, the building of which can destroy natural settings.

Although power boats probably have the greatest impact overall, much depends on the way in which particular activities are carried out. This also holds for the social sustainability of activities, with motorized forms often being criticized for interfering with the enjoyment of others and sometimes even endangering them (Pigram, 2003; Richins, 2007). Efforts are being made, however, to manufacture marine engines that emit less hydrocarbons and less noise.

In terms of encouraging sustainable behaviour, many organizations and governing bodies have developed guidelines and codes of conduct for sensitive participation. FISA's Environment Working Group, for example, suggests that rowers reduce noise and boat wake so as not to disturb sensitive habitats, particularly during nesting and spawning seasons, avoid shoreline erosion and not discard any waste into the water. They also advise that powered support craft should minimize pollution and take care not to spill fuel or oil into the water (http.//www.world rowing.com/medias/docs/media_354730.pdf, retrieved 9 May 2010). For their part, the CMAS direct the *sub aqua* NCAs to avoid contact with underwater plants and corals, and not to stir up sediment or disturb wildlife (http://www.cmas.org/114-29577.php?commission=). The International Waterski and Wakeboard Federation has compiled an extensive Environmental Handbook which recommends participants maintain a minimum distance of 50 m from the shore to mitigate erosion and impacts on wildlife, and a minimum depth of 2 m to reduce turbidity (see www.iwsf.com).

Among the groups of NCA participants taking an advocacy role, open water swimmers have been involved in promoting clean rivers and lakes and raising public awareness about water quality. The IRF promotes clean, accessible free-flowing rivers for rafting and organizes river clean ups to collect rubbish. Rafters generally have been involved in opposing damming, diversion, dredging and other modifications to natural river systems (L. Jones, 2007). One of the most successful and visible groups campaigning on environmental issues has been surfers – both the UK-based Surfers against Sewerage (SAS) and the US-based Surf Foundation, which has over 50,000 members and 90 chapters worldwide (http://www.surfrider.org/whoweare.asp, retrieved 29 May 2010). The wide ranging concerns of SAS include sewage, marine litter, climate change, toxic chemicals, shipping and protecting the recreational wave resource and they aim to 'promote a clean and safe water environment for people to enjoy and practice a wide variety of sports that rely on these marine resources' (http://www.sas.org.uk/sas-background/who-we-are/, retrieved 29 May 2010).

With the water NCAs we have, potentially, some of the least consumptive options to engage in a nature challenge, particularly with open water swimming, snorkelling and freediving. But while the greatest extravagances may be associated with luxury cruising and power boating, within each activity consumption – like environmental sustainability – can vary widely depending on patterns of behaviour. Snorkellers and freedivers may travel extensively to prime locations, while scuba divers could spend more on equipment, but be content to explore their local dive spots and save on travel expenses. Similarly, although participation in the surf NCAs requires relatively little in the way of basic equipment, these activities support a lucrative specialized travel industry, as well as large multi-national corporations producing brand-name merchandise such as clothing, sunglasses and watches which are readily consumed by participants eager to establish their identity and status within an image-conscious social world.

4
Land

Land, as a natural element, may be encountered as various formations and consistency of rock, sand and earth – for example, mud and clay – accompanied by associated vegetation and within settings that are just as various, from airy alpine heights to dark, intimate subterranean passages and pleasant forested paths. Within this breadth and diversity, *homo otiosus* has found a plethora of natural challenges and increasingly specialized ways of enjoying them. Land-based NCAs can be self-propelled, as we find with our first group of activities discussed in this chapter, beginning with hiking which is perhaps the most accessible NCA of all, both geographically and economically. Alongside hiking, we have the kindred activities of multi-day hiking and wilderness camping, mountain and trail running and scrambling. Also among the self-propelled NCAs pursued on land are the more technical variants of mountaineering, rock and ice climbing, canyoning and caving. Mountain biking follows, then orienteering and geocaching which combine elements of the preceding core activities with an additional navigational challenge. A brief mention is made of wind-propelled land-based NCAs, before we discuss wilderness horseback riding, off-road motorcycling and 4WD.

Before we begin an explanation is required regarding a number of activities we have classified as land-based, but which can also involve encounters with water, ice and snow. While mountaineering occurs on diverse terrain including ice and snow, and could therefore have featured in Chapter 6, we included it here because we found it more logical to discuss it alongside the closely related activities of hiking, rock climbing, scrambling, caving and so on. Ice climbing, as a specialized form of mountaineering, is also covered here because the techniques and equipment used are so akin to general mountaineering and rock climbing that it seemed a more obvious fit in this chapter. Nonetheless, where land-based NCAs occur in cold and wintry environs, many of the attractions and challenges discussed in Chapter 6 also come into play. Likewise, NCA enthusiasts negotiating canyons and caves can be faced with not only a variety of land features, but also the element of water, and hence these activities overlap with those covered in Chapter 3.

Hiking

Hiking is the core activity from which a variety of land-based NCAs extend and is also the means of accessing remote areas in which a range of NCAs are undertaken. Also known as hill or bush walking, backpacking, trekking and tramping, hiking is essentially walking for pleasure in natural surroundings. In its most basic form very little experience or special skill and equipment is required, beyond what comes naturally to us and the acquisition of suitably sturdy footwear. But from this modest entry level, a hiker may progress by increasing the degree of difficulty, strenuousness and/or length of trip undertaken in order to attain the intensity of challenge desired. This 'natural progression' may in time lead them into the related activities of scrambling, rock climbing, mountaineering, canyoning and so on. The appeal of walking in backcountry areas, whether on a simple day walk or a multi-day wilderness camping trip, is to enjoy the natural surroundings, free from the intrusion of the clutter and noise of 'civilisation'. A particular attraction is often the escape from motorized forms of transport and the ability to access areas that cannot be reached by any other means. On more extended and difficult hikes, satisfaction is found in achieving the level of physical fitness and endurance required, and attaining a degree of self-reliance by which one can survive in and appreciate unmodified wilderness areas.

Indeed, the less technical nature of hiking in relation to some other NCAs allows more opportunity for the quiet contemplation of the natural surroundings and, for many, a sense of connection with nature, as described by an American hiker:

> It's that reaction most of us experience watching the sun rise over a lake or standing on a ridge looking down over a timeless, beautiful canyon. It's that deep connection with landscape that sometimes overwhelms us during a long hike, where nature is clearly more than an amenity – it's a basic human need, embedded in our genetic code.

(G. A. Miller, 2007, p. 6)

Walking as a recreational activity in natural settings began in the late eighteenth century in Europe, strongly influenced by Romanticism, and gained widespread popularity during the twentieth century with increasing leisure time and improvements in transportation. It gained particular momentum in the inter-war period in North America, Western Europe, Great Britain, New Zealand and Australia, with the establishment of clubs (Harper, 2007, p. 169; Ross, 2008). While hiking cultures in different countries acquired their own specific cultural characteristics, they were often linked to the embracing of nature and the rejection of the ills of modern life, and led to the growth of social and political movements concerned with the protection of wilderness.

Suitable terrain for hiking includes forests, mountains, deserts and coastal regions. While the majority of hikers follow pre-established trails, more experienced hikers may prefer to determine their own routes across unmarked ground using their navigational skills. Hikers carry with them all the necessities for survival, for example food, water, adequate clothing, first aid and emergency equipment, depending on the terrain and conditions to be encountered.

What is the nature challenge? Although hiking in its most basic form requires no special skill, particularly on well-formed trails, physical fitness is required on steep and/or arduous ground, and strength and endurance is required for long hikes. Under such conditions, it is advisable to walk at a steady, rhythmical pace in order to avoid exhaustion. When ascending steep slopes, short steps at a slow but sustained pace conserve energy, while placing feet as flat as possible (e.g., by stepping on rocks and tufts of grass or following a zigzag pattern) will ease the strain on the calf muscles. Off-trail, or on poorly formed trails, the hiker must develop surefootedness and balance, with particular skill required in negotiating uneven ground, unstable steep surfaces (e.g., scree), crossing rivers, snow and ice if encountered. While for the novice or the unfit hiker, a hike may be more akin to a plod, the ability to traverse difficult terrain on foot in a competent, smooth and efficient manner can engender a sense of flow. When off-trail, the additional skill of route-finding may be required to identify a route across a landscape that avoids natural obstacles such as cliffs, dense vegetation, soft sand or bogs and is therefore the least arduous, dangerous or technically demanding. Hikers must also be prepared to deal with adverse weather conditions, from intense heat to electrical storms, torrential rain and blizzards. Depending on the nature of the hike they may need to be able to navigate using map and compass or a GPS. In particular locations wildlife may pose a challenge which hikers need special knowledge and skills to negotiate. This could be potentially life-threatening, as in the case of bears or snakes, or merely unpleasant and painful, such as mosquitoes and leeches.

The combination of the core challenges and the great variety of terrain and conditions that may be found while hiking give the activity its special character, as summarized here by a New Zealand 'tramper'.

> Tramping is an immediate and very real life. At times, a life that can be laid back...at other times, a life that can demand every fibre of effort, and be laced with much discomfort. Sleeping in a wet sleeping bag on a cold wet night; fighting wind and sleet on the ridges; sleeping in the sun on river flats: it is just this variety that gives tramping its life.
>
> (Spearpoint, 1985, p. 13)

In Australia, 'bushwalking' is the only NCA among the top ten physical activities in terms of annual participation. Having experienced a 34 per cent increase in participants since 2001, 6.4 per cent of the population, or 1.1 million people went bushwalking in 2008 (Australian Sports Commission, 2008). It is also the most popular NCA in the United States, with 11.6 per cent of the population, or 32.5 million people hiking in 2009, representing a 8.4 per cent increase over the preceding decade (Outdoor Foundation, 2009a, 2010).

All of the self-propelled, land-based NCAs that follow, with the exception of mountain biking, share common elements with hiking, particularly wilderness camping, mountain and trail running, scrambling, orienteering, geocaching and less technical forms of canyoning. The more technical pursuits of climbing and caving are frequently preceded or interspersed with spells of hiking, so it is hard to escape this most basic of the land-based NCAs if one heads across country in search of a challenge.

Wilderness camping and multi-day hiking

Wilderness camping and multi-day hiking combine hiking with overnight stays in the backcountry, with trip length being anything from a weekend to weeks and months. This extension of hiking allows participants to access more remote areas and spend more time in undisturbed natural settings, and may be combined with other NCAs, such as fishing, hunting, nature study and photography. In some areas basic huts may be available to provide shelter, warmth and sometimes cooking facilities for hikers. Otherwise, hikers intending to camp in the wilderness must carry with them all necessary equipment to meet their needs for food and shelter, including cookers and pots, sleeping bags and a tent. Elsewhere hikers may be able to erect simple shelters using natural materials, or find a suitable bivouac under an overhanging rock or in a cave.

Studies have found that wilderness camping is most passionately carried out by people brought up in cities and who had early camping experiences (Meyersohn, 1970). The extended nature of this form of NCA allows for sustained absorption and interaction with the natural environment, a related heightened sense of escaping from urbanization and the challenge of exploring and surviving in the backcountry which leads to satisfying feelings of self-reliance. This self-reliance is particularly enhanced on solo trips, while travelling with groups over extended periods can intensify the degree of sociability and develop strong bonds between hiking companions.

Multi-day trips take place in backcountry or wilderness environments which are ideally large enough to enable travel on foot for more than a day without encountering urban development. In addition to the appeal and core challenges of hiking, multi-day hikers must have the necessary strength and endurance to carry a backpack over a long distance. Knowledge and experience contribute to important decisions regarding what to take and what to leave behind in order to reduce unnecessary weight. Skill is also required to locate suitable camp sites which are sheltered from the elements and free of potential natural hazards such as flooding rivers and rock fall. Backcountry cooking is an additional challenge on overnight trips, the test being to produce as tasty and nourishing a meal as possible from light weight, transportable ingredients and requiring only the sort of preparation that can easily be met on a camp cooker or open fire.

Long-distance backpacker Chris Townsend describes the particular challenges and rewards of his preferred NCA:

> A journey on foot is the best and arguably only way to journey through a landscape and really see it...Once the journey is well underway and the little niggles of the first days – the concerns over equipment, camp sites, water sources and route finding – have faded away its nature changes. The experience becomes deeper and more intense and I can concentrate on the land and the walking and camping. On multi-week hikes it becomes my way of life. This is what I do, this is what I am.... Backpacking, slow and inefficient at

getting anywhere, is about the process not the product, about enjoying nature and land, about relishing the physical effort of walking and the skills of navigation, camping and coping with the terrain and the weather.... Backpacking is the finest way to lose yourself in a landscape. By spending days or weeks moving though the land and sleeping there at night you become attuned to its characteristics (that soft down-slope wind that arises after dusk, the bright green attractive-looking ground that signifies a deep bog), its smells, its plants, its wildlife, its feel. And as you move through the landscape you can watch it change, watch mountains and rivers grow and diminish, watch forests deepen in the valleys and thin and dwindle as you climb, watch the shape of the land gently alter as rocky peaks give way to rounded hills and the latter in turn to low moorland or forest.

(Townsend, 2009, pp. 78–79)

Neither hiking nor wilderness camping are practised as a sport. However, adventure racing (discussed in Chapter 7) incorporates aspects of these activities in a competitive event. In the United States, 7.8 million people went overnight backpacking in 2008 (Outdoor Foundation, 2009a).

Mountain and trail running

Also known as fell or hill running, mountain/trail running is the recreational or competitive activity of off-road running on steep terrain. The appeal of the activity over its road-based counterpart is the enjoyment of natural surroundings, and the challenge of negotiating rough country with demanding ascents and descents.

Mountain/trail running can take place anywhere from alpine regions to hills no higher than 200 m, and ranging in difficulty and length from 15-minute sprints to several hours. Participants normally utilize established walking paths, often carrying with them small backpacks containing water, food and extra clothing as required according to the prevailing climatic conditions. The surface underfoot can vary from mud to sand or clay, with rocks, tree roots and streams to negotiate. Although the setting is the same as for hikers, the emphasis for mountain/trail runners is on fitness and endurance, with satisfaction coming from covering more ground more quickly. Ascending and descending steep terrain as quickly as possible requires particular muscular strength and endurance and cardiovascular efficiency. Nonetheless, mountain and trail running share with hiking the central attraction of accessing remote areas of natural beauty, as illustrated by the following account from a 'fell runner' in the United Kingdom.

> I savour a doze in the hot sunshine, before resuming my journey with the climb of Ladhar Bheinn. The 1000 metre ascent from sea level following the previous three big climbs and descents is almost too much for me. The stalkers' track weaves in and out of chest deep bracken, before traversing out to Coire Dhorrcaill and emerging on to yet another very steep grassy slope. My progress slows to a crawl, but I am committed and can only drag myself onwards. The superb cliffs of Ladhar Bheinn act as a temporary diversion, but even searching for future winter routes fails to relieve the omnipresent effort of the upward grind. This is self imposed torture with no-one to cry to, no-one to shout at, no-one to blame. Why am I doing this? The answer comes, eventually, on the summit of Ladhar Bheinn where the expansive view of the Western Isles and ridges to the East lift me out of my torpor. The breeze revives me and I relish the privilege of my solitary perch. I have seen no-one since the start and only meet three parties a little later in the day.
>
> (http://www.fellrunner.org.uk/articles/0410/
> roughride.pdf, retrieved 14 January 2009)

Like their on-road cousins, mountain and trail running are often pursued competitively. Early organized hill runs took place in the United Kingdom as part of community fairs and games from the mid-nineteenth century, with cash prizes for winners. From the mid-twentieth century, mountain/trail running developed as an amateur endurance sport. The World Mountain Running Association has staged a World Mountain Running Trophy since 1985 and in 2003 the event attracted entries from more than 30 countries. Longer races may also require competitors to utilize navigational skills, and thus mountain running overlaps with forms of orienteering, particularly rogaining, discussed later in this chapter. Trail running is another NCA on the rise in the United States, with a 16 per cent increase in participation over the past decade, and a total of 4.8 million participants in 2009 (Outdoor Foundation, 2009a, 2010).

Scrambling

Scrambling, or alpine scrambling, is the term used to describe activities which fall somewhere between hiking and rock climbing, and where

the goal is generally to reach a summit which does not require technical climbing skills and equipment. Scrambling is typically distinguished
from hiking by the use of one's hands to negotiate off-trail terrain.
The nature of this terrain may involve vegetation, rock and possibly snow where alpine scrambling is involved. The nature challenge
of alpine scrambling is described on The Mountaineers website as
'negotiating lower angle rock, travelling through talus and scree, crossing streams, fighting one's way through dense brush and walking on
snow-covered slopes' (http://www.mountaineers.org/seattle/scramble/
FAQ.htm, retrieved 24 October 2009). While scramblers may, according to their level of experience, use a rope on certain sections of difficult
and/or exposed routes, one of the attractions of scrambling is the freedom of movement which comes from being unencumbered by technical
climbing equipment. Various grading systems are used to indicate the
level of challenge presented by a particular scramble, taking into consideration the difficulty of route finding involved, the level of exposure
encountered and the degree of technical climbing skill and rope use
required.

As with hiking, the satisfaction of this NCA is to be found in the
fluid and efficient movement over mixed terrain, in this case combining
secure footing with the use of one's hands for steady balance and additional traction on steep ground. Branch Whitney describes a day out
scrambling in Nevada.

> As I drove into Red Rock Canyon, I knew I found paradise. From bril-
> liant red sandstone foothills to massive 3,000-foot sandstone cliffs,
> I could hardly wait .... Scrambling up the multi-colored wash was a
> blast. I'm glad I am in good shape, because in less than a half-mile
> I scrambled over 1,000-feet to the ridgeline. At the top the terrain
> changes from the typical limestone to sandstone. And I mean acres
> of sandstone!
>
> I followed the book's directions and started traversing around Ice
> Box Canyon. The views along the traverse were fantastic. The floor of
> Ice Box Canyon must lie at least 1,000 feet below .... Finally, I got my
> first look at Bridge Mountain. It's huge dome-shape peak that looks
> like it doesn't want to be climbed.
>
> As I started the 'traditional' approach to Bridge, there were cairns
> and black slash marks to help guide me. The route takes you down
> a couple class III chutes before reaching the bench that goes over to
> the base of Bridge Mountain. Now, from the bench, Bridge Mountain
> looks like a technical climb. The book reassures readers that it's only a

class III climb – I wasn't so sure. As I got closer, the crack that looked so intimidating began to look doable.... It was a thigh-burner and it did have some exposure. The views from the peak took my breath away. All of Red Rock Canyon laid in front of me.

(http://classic.mountainzone.com/hike/pub/stories/
02-05-99.html, retrieved 16 January 2009)

Mountaineering

Where the primary destination of a land-based NCA is a mountain that requires specialized climbing techniques and equipment, hiking and scrambling give way to mountaineering. The specific techniques required depend on the terrain to be traversed, and general mountaineering requires proficiency on rock, snow and ice.

While there have been instances of people climbing mountains for a variety of reasons throughout recorded history, histories of mountaineering as a leisure pursuit in its current manifestation tend to begin with the English Victorians, who founded the first club for mountaineers, the Alpine Club, in 1857 (Lunn, 1957). An interest in climbing mountains for essentially recreational purposes was awakened in European minds over the preceding centuries by the Scientific Revolution and rise of the Romantic tradition which, according to MacFarlane (2004), allowed the mountains to be 'imagined' differently from before: as geological formations and places in which to encounter the sublime, rather than the lairs of dragons and the haunts of devils. During the eighteenth century, a small number of individuals were climbing in the European Alps and, by 1850, a few major ascents had been made by continental mountaineers. In the latter half of the nineteenth century, the English came to the forefront of mountaineering in Europe. By 1880, having climbed all the major peaks in the Alps, the English and their continental guides took their techniques and experience and launched expeditions to mountain ranges in other parts of the world (Davidson, 2002; Temple, 1969). Mountaineers can now be found practising their craft throughout the world, wherever mountain ranges and alpine environments are to be found.

The nature challenge of mountaineering involves using a range of skills and equipment to traverse and ascend the varied terrain found in alpine environments. When traversing hard snow or ice, specialist equipment includes crampons, which have sharp points and are attached to the bottom of a mountaineer's boots, thereby gripping the hard surface underfoot. On soft snow, snowshoes or skis may be worn (see Chapter 6). When ascending or descending steep snow slopes, an ice axe is carried. Ice axes have evolved into specialized forms, but for general mountaineering they have an adze which can be used for chopping steps in hard snow or ice and a head with a slightly curved pick used for self-arrest: that is, when a mountaineer falls on a snow slope the axe is held close to the body and the head is driven into the snow to slow and/or break the ensuing slide. The shaft of the ice axe has a spike at the bottom which provides grip when it is held by the head and used like a walking stick. The length of the ice axe depends on the nature of the terrain, with a shorter axe being suitable for steep slopes and ice climbing, while a longer axe is of more use as a walking stick, when crossing glaciers for example. On steep slopes the pick of the ice axe can be driven into snow or ice to provide grip.

Moving smoothly and efficiently is essential to good mountaineering. Speed is important in order to avoid adverse conditions such as soft, unstable snow late in the day, the onset of bad weather or night fall. However, moving in a rushed and haphazard manner can lead to mistakes. Solid footwork on varied gradients is critical, from kicking flat steps where possible into a snow slope to harder surfaces on which crampons are worn and utilized by either placing the foot on the slope so that as many of the points as possible have contact with the snow (known as flat-footing or 'French technique' and suitable for lower-angled slopes) or by 'front pointing' on steeper and harder-surfaced slopes (see below), or a combination of the two techniques. Ropes may be used in conjunction with a variety of different 'anchors' to secure mountaineers to the mountain and stop them in the event of a fall. Anchors include snow stakes, which may be driven into or buried in the snow, ice axes, ice screws, skis and rocks. Anchors may also be created by carving or drilling suitable indentations in hard snow or ice. More specialized devices have also been developed for creating anchors in rock features, discussed in the following section on rock climbing.

Mountaineering also requires navigational and route-finding skills, the ability to read weather patterns and often coping with the effects of altitude, which can include dehydration, drowsiness, headache, loss of appetite, shortness of breath and may progress to life-threatening altitude sickness. Other hazards that mountaineers must successfully negotiate include avalanches, ice and rock fall, crevasses and adverse weather which may trap them on a mountain for extended periods. The severe cold often encountered in alpine regions can also lead to hypothermia and frostbite unless mountaineers have adequate clothing and shelter.

In her biographical narrative study of committed mountaineers, Davidson (2006) found that long-term involvement in this NCA was underpinned by a deep love and feeling of connection to the mountains, which often grew out of childhood experiences. Despite the activity's reputation for thrill and adventure, mountaineers frequently spoke of a sense of calm, peace and being 'at home' in alpine environments. The challenges of climbing mountains, and the self-reliance this demands, gave the mountaineers a sense of being able to discover and express their 'true selves', while the camaraderie of the climbing community nurtured a set of shared values which helped to sustain their commitment to this NCA.

While mountaineering itself has not been formalized as a competitive sport, the specializations that follow (rock and ice climbing) have developed organized competitions.

Rock and ice climbing

While techniques for climbing rock and ice emerged at the same time as mountaineering, both have since also developed as independent specialized forms of climbing and NCAs in their own right. The origin of rock climbing as a separate activity is traced to the latter half of the nineteenth century, spurred on by mountaineers looking for ways of developing skills and fitness during the alpine offseason. Although distinctive ice routes were being established around the same time, it took longer for ice climbing to achieve a similar status as an independent activity, as it awaited the development of front points for crampons in the 1930s and specialized ice tools in the 1960s.

As specialized climbing activities, ice and rock climbing are generally undertaken on rock or ice features where the objective is to complete a defined 'route', rather than necessarily to ascend to a mountain summit. Where such routes are found in alpine environments, the lines between rock/ice climbing and mountaineering are blurred. The distinction may be one of access. Rock and ice routes in mountainous areas require general mountaineering skills to reach, whereas more easily accessible routes can be found near a car park or at the end of a short hike, thereby allowing the specialist to focus on the technical aspects of their climbing without the need to attend to the many other aspects of travelling in the mountains, such as navigation, avalanche safety, changeable weather and so on.

Rock climbing can in itself be differentiated as traditional (or adventure) climbing, sport climbing and bouldering. The core activity of rock climbing involves utilizing the natural features of the rock, such as cracks, grooves, ledges, slabs and 'chimneys', to move up the route, with various techniques particular to the different types of rock and their forms. The nature challenge posed by this environment requires the careful selection of suitable hand holds, foot placements and body positions to allow upward movement as efficiently as possible. The distribution of weight and balance is crucial to good technique, as are strength and agility. Typically climbers wear specialized rock shoes with high-friction rubber soles which help them to find adequate foot placements, allowing body weight to be supported by the legs and relieving stress on the arms and fingers. While brute force may propel a climber to success on a chosen route, rock climbing is more intrinsically satisfying when fluid and well executed moves make the climb as efficient and 'dance-like' as possible.

Unless the climber is 'soloing' without the protection of a rope, then he or she must also master a technological system of ropes, karabiners and 'protection'. Adventure or traditional climbing requires the placement of various devices collectively known as protection in fissures or 'points of weakness' within the rock by the 'lead' climber while they are ascending the rock face. The climber is tied into the end of a rope, which is run through the protection as it is placed. The other end of the rope is attached to the 'second' or 'belayer' below, who feeds the rope through a friction device which will hold the rope taut should the climber fall. This technique of placing protection means that the natural features of the rock largely determine the direction of the route, along with the climber's skill in route finding and the skilful placement of adequate protection. An additional challenge for the climber is to control his or her fear, which may be compounded by the degree of exposure on a route, and the degree of confidence in the protection that has been placed. The combination of these various challenges requires deep concentration which facilitates the feeling of flow (Macaloon & Csikszentmihalyi, 1983) and, particularly when set within a stunning natural setting, leads to very memorable experiences.

The satisfying combination of natural challenge in an awe-inspiring setting is recalled by a New Zealander climbing in Montana with her close friend Lisa.

> ...it was due to storm that day so we did this climb, it was five pitches that you could get off quite easy. And, it was graded 5.11, it had four pitches of 5.10, then this hard 5.11 pitch,[1] and it just went real smoothly and we had a wicked time. It was up this valley that

didn't have any paths, no one was around, just these beautiful sort of wooded hills, but Montana feels real big, real vast, in a way. We could just see the storm clouds rolling in, this chilly breeze and things, and we got to the base of this 5.11 pitch and it, just looked, like nails-hard. And we'd both been thinking during the day, 'oh well we'll see how we go. We can always back off.' And it was this flarey, kind of finger crack...just those moves where...you're somehow staying on, but you're not quite sure how, and [you] just sort of keep going another move...And Lisa's sort of yelling, I'm grunting, got this massive pump going, yeah [I] just kept on going to this rest, and...I'm always torn, like I really want to have finished it, so that you've done it, but then another part of me never wants to finish, wants it to kind of keep going....it's just the most amazing feeling of, just, I don't know...it's not like you're battling the rock or, any sort of thing like that but, you're battling yourself I guess, just to stay on. Yeah, I love that [laughs].... It was a magical day.

(Davidson, 2006, p. 110)

With sport climbing, which depends on pre-placed fixed bolts to provide protection, routes can be established and climbed where the condition or characteristics of the rock would make it otherwise difficult to protect. For the sport climber there is greater emphasis on the technical difficulty of the route without the concern for having to place protection and determine the correct route. Lewis (2004) argues that this distinction fundamentally alters the embodied experience of climbing with the sport climber being alienated from the environment by their reliance on technology which overrides the natural challenges of the rock.[2] Where the placement of technical devices into the rock is also used to provide extra hand and footholds, this is termed 'aid' or 'artificial' climbing.

Bouldering involves ascending boulders or small outcrops, and boulderers rarely go more than a few meters above the ground, meaning that the use of a rope is unnecessary. Instead a boulderer may use a crash pad or a 'spotter' to help protect them in the event of falling or jumping off. The focus of bouldering is on technique and problem-solving applied to individual moves or a short series of moves which require strength and dynamism, rather than the mastering of a technological system including ropes, protection, belays and so on. Bouldering has been increasing in popularity since the late 1960s, with a surge in participation and worldwide appeal during the 1990s (http://www128.pair.com/r3d4k7/Bouldering_History1.0.html, retrieved 2 December 2009).

World renowned bouldering sites include Fountainbleu in France and Hueco Tanks in the United States.

Ice climbing as a specialized activity generally takes place on frozen waterfalls or flows over natural outcrops. Being more ephemeral than rock, the consistency and form of the ice can vary greatly, and this contributes to the natural challenge that the activity poses. Ice climbers use equipment designed specifically for the purpose – ice tools, similar to axes but generally with a curved shaft and more sharply angled pick, and specialized rigid crampons suited to 'front-pointing' up vertical ice. The basic movement up the ice route is achieved by kicking the front points of the crampons into the ice and stepping up on them directly. The climber then reaches above his or her head with the ice tools held by the base of the shaft and swings them successively so that the picks penetrate the ice. These movements are repeated in a rhythmic and balanced way. Rope and protection techniques are similar to rock climbing, with the main protective device used being an ice screw.

Mixed climbing is a recent hybrid of rock and ice climbing which evolved from the desire to reach ice features which began high off the ground, and to connect sections of ice separated by patches of rock. By using a technique known as 'dry tooling', ice climbers scale rock using their ice tools and crampons, placing them delicately on rock features, thus eliminating the need to switch equipment when moving up mixed ground.

Sport climbing, bouldering and ice climbing can be further differentiated from mountaineering and traditional rock climbing as they can be practised in artificial environments, primarily for training and the organized competitions which have been established around them. Here natural features are simulated, but under these conditions sport and competition climbing cannot strictly be considered as NCAs.

The 2009 figures from the United States indicate that 1.8 million people participated that year in either traditional/ice climbing or mountaineering, while 4.3 million went sport/indoor climbing or bouldering (Outdoor Foundation, 2010). In the United Kingdom in 2008/09 the monthly participation figure was 228,200 for all types of climbing and mountaineering (Sport England, 2009).

Canyoning

Canyoning, also known as canyoneering (US), kloofing (South Africa) and river tracing (Japan), involves travelling through or across canyons, ravines or gorges using various means ranging in degrees of technicality from hiking, boulder hopping, jumping, swimming and wading to abseiling, scrambling and rock climbing. In addition to these technical aspects, it can also require navigational and route-finding skills. Equipment used by canyoners includes Personal Flotation Devices (PDFs), wet suits, specialized footwear with high-grip rubber soles and other general climbing equipment (ropes, helmets, harness, slings etc.).

A closely related activity is coasteering, where participants follow a coastline rather than a water course. Similar techniques are required, with additional natural challenges posed by waves, tides and currents, estuaries, lagoons, sand dunes and seaweed, and possibly even sharks. Like canyoning, coasteering requires warm clothing, wet suits and PDFs, and hard-wearing non-slip shoes. Also advisable are trekking poles for balance on slippery rocks and a knife to free oneself from kelp in case of entanglement.

It appears that people have been travelling through canyons for recreation for at least as long as they have been hiking. Significant developments were made in canyoning in the 1960s alongside the evolution in equipment and techniques for mountaineering and rock climbing. Canyoning slowly gained a following in subsequent decades in the United States, Europe and Australia, with its popularity growing throughout the 1990s and an increasing number of new routes being established around the world, including China, South America, Hawaii, Thailand and Africa. Canyoning has since become a distinct activity with increasingly specialized equipment manufactured exclusively for use by canyoners. Despite clear similarities with other land-based NCAs, particularly hiking and rock climbing, canyoners can be insistent that their activity has a unique nature and identity. As the website of an American alpine adventure training company claims: 'The methods, mindsets, and equipment needed to safely descend a technical canyon route are far from any systems that a rock climber would ever use' (http://www.alpinets.com/canyoneeringhistory.html, retrieved 9 December 2008).

The unique natural landscape of canyons and the range of technical challenges they pose, including both rock and water features, clearly distinguish canyoning from other land-based NCAs. The activity may be undertaken anywhere that offers suitable canyons and ravines. Those with narrow passages, frequent drops, waterfalls and sculptured rock walls (sandstone, granite, limestone and basalt predominantly) provide satisfying challenges and scenic appeal. The level of challenge may range from a straightforward hike to a difficult and technical route requiring competent rope work, climbing and descending skills. A hiking manual informs prospective canyoners that they should: 'Expect to swim and wade through icy water, negotiate rapids and waterfalls, jump off cliffs into deep pools, abseil down canyon walls, scramble or climb along steep cliffs and jump from rock to rock across gaping chasms' (Marais, 2002, p. 64).

American canyoner, Malia McIlvenna, describes her 'obsession' with the activity:

> It didn't creep or slowly develop, but rather, from the moment I first heard of canyoneering, my obsession was entire and immediate. The remoteness, rugged beauty, adrenaline, excitement, wonder, challenges and intimacy of the sport have taken over my life. Hardly a day passes when I don't dream or at least talk of the canyons which fill my heart.... The destination is only my reward – the work of getting there is all the fun! I enjoy the challenge of orienteering, thrill of discovery, fear of the unknown, wonder and awe at the wildlife I encounter, burning muscles as I scramble up steep slickrock, climb over boulders and squeeze through slots, and the blissful peace of solitude.
>
> (http://www.math.utah.edu/~sfolias/canyontales/tale/? i=obsession, retrieved 9 December 2008)

Aside from the core activity, another layer of challenge for the canyoner is the level of commitment involved in entering a canyon. Once *en route*, it may be difficult to escape from a deep canyon with sheer sides should an emergency arise. Canyoners must take precautions to avoid hyperthermia in hot, dry canyons and hypothermia in dark, wet ones. The presence of water also poses particular challenges including deep water and submerged obstacles when wading or jumping into pools. But perhaps the greatest potential hazard for canyoners is the flash flood, and canyoners must be adept at reading weather forecasts, estimating the extent of the watershed that feeds into a canyon and watching carefully for signs that flooding may occur. Unwary canyoners have been caught out by underestimating the possibility of flooding, particularly as localized rainfall does not necessarily indicate the level of precipitation over a whole water catchment which can be channelled into a canyon with great speed and ferocity. Through needing to understand and respect such forces of nature canyoners, like other NCA enthusiasts, develop a special relationship with the landscapes they traverse. As one canyoner puts it: 'Canyoneering brings us ever closer to nature, reminds us of our humanity, celebrates the clarity of living in the moment' (http://www.alpinets.com/canyoneeringhistory.html, retrieved 9 December 2008).

Caving

Caving has much in common with canyoning in its combination of negotiating rock and water, and with its use of general climbing techniques and equipment adapted to suit the special nature of a distinctive subterranean landscape. But the dark, other-worldly and frequently spectacular haunts of the caver offer a very particular challenge which may horrify many, but fascinates and tantalizes a dedicated few.

Caving's peculiarity is described by the New Zealander Carl Walrond.

> If you're even slightly claustrophobic, it's the stuff of nightmares: you're deep underground and it's pitch black, cold, and sometimes too narrow to turn around. But caving attracts adventurers who are unfazed by all this. In their quest for new passageways, they are drawn by the eerie beauty of cathedral-like caves, waterfalls, delicate rock formations and shimmering glow-worms.
>
> (http://www.teara.govt.nz/TheBush/BushAndMountain Recreation/Caving/en, retrieved 14 December 2008)

Like many of our land-based NCAs, recreational caving's origins are traced to the second half of the nineteenth century, with the exploration of caves in Europe and the United Kingdom (Cant, 2003; Taylor, 1996). As with early mountaineering, the beginnings of the hobby are intertwined with exploration and scientific discovery. Even now, cave systems represent some of the last unexplored and most mysterious regions remaining on earth and the discovery and surveying of previous unknown cave systems can be a strong attraction of the activity.

Cave systems exist all over the world, where they have been fashioned by the natural processes of water, waves, glaciers and lava. Particularly common are caves formed in limestone or marble, which is easily eroded by seeping water which has absorbed carbon dioxide from the air and thereby become mildly acidic. This slow action over vast periods of geological time results in underground caverns, tunnels and potholes (vertical shafts) which can stretch for many kilometres. In the hollow spaces created, fragile speleothems such as stalactites and stalagmites form and specially adapted life forms live. Because of the unique conditions under which cave environments have existed they are extremely vulnerable to destruction by visitors from above ground.

A careless nudge can obliterate a crystal latticework thousands of years old. The oil from the touch of a human hand can permanently halt the growth of a forty-foot stalactite. Even the layered mud of a virgin floor is noticeably changed by a single caver's crossing.

(Taylor, 1996, p. 13)

The clothes worn and equipment used by cavers depend on the specific features of the cave being entered. In cold caves where water is present, a full wet suit may be worn underneath a hard-wearing set of coveralls designed to protect the caver from abrasive rock when crawling or sliding through narrow slots and passages. In hot, dry caves thin polypropylene clothing is often worn to keep as cool as possible, while also providing some protection from rock surfaces. Other protective gear includes sturdy boots, knee-pads and elbow pads. A hard hat protects the head from falling rocks and bumps in tight spots. Of crucial importance of course is a light source which is generally mounted on the helmet, with two alternative sources of light typically carried as a contingency against failure, damage or loss.

The terrain of a cave may necessitate vertical ascents and descents using rope systems. Cavers may also need to squeeze through narrow slots, crawl along confined passages and scramble over jumbled boulders in the semi-darkness.

A novice caver describes the sensations she experienced on one of her first trips into a cave.

Each foot is placed deliberately, the next move already determined. Debbie leads me, calling out directions, showing me how to hoist my body, how to use my knees, how to lean into a boulder and inch sideways. Constantly aware of the fragility of my body, I work these stones like a slow motion, 3-D hopscotch, searching out the safe foothold, the wide-enough ledge, the handhold that will keep me from falling. I forget that I am deep inside a mountain. I forget about everything but the next move. And the next and the next and a crawling, scrambling, exhilarating, exhausting hour later, we have worked our way through. My legs are trembly, my knees ache, and I'm sure that underneath my overalls my skin has begun to bloom in bruises.

(Hurd, 2003, p. 10)

Despite its commonalities with climbing, including techniques and equipment, according to Cant (2003, p. 70) 'caving draws upon a slightly different set of embodied-sensuous and aesthetic experiences to its kindred sport of climbing, that relate directly to physical spaces unlike any found above ground'. This involves 'shifting senses of enclosure and space' as well as the defining element of darkness which can evoke powerful emotions and ensures that the usually dominant visual sense is overridden by other senses. The dark can also subtly alter the social experience of caving, as Hurd (2003, p. 160) explains:

> Everybody enters a cave dressed in rough overalls and hard hats, boots with good tread and gloves. You can study the others, try to get a sense of body shape and maybe age by limberness or lack of it. But that's about it. You could be inching on your behind down scrabbly slope in the mostly-dark next to the Queen of England and you'd never know it. The only thing that matters in a cave, they told me, is your ability to stay calm in dark spaces.

In a squeeze, the challenge for cavers is to use the limited movement of their feet and hands to continue forward movement and to twist their torsos to avoid becoming wedged or paralysed by panic. In extremely tight spots, forward movement may only be possible when a breath is exhaled. Notorious squeezes are given suitably daunting names: the Agony, the Back Scratcher, the Claustrophobia Crawl, the Grim Crawl of Death, the Devil's Pinch. When a stream is running through a crawl space, the caver must also ensure that adequate breathing space is allowed for. In completely submerged passages and pools, cave diving may be practised using scuba gear (see Chapter 2). Cave diving is a highly specialized and relatively dangerous off shoot of caving (Taylor, 1996).

While there are a proliferation of caving clubs and national organizations, caving, like mountaineering, is not practised as a competitive sport. Monthly participation in caving in the United Kingdom was 9900 in 2008/09, up from 5100 in 2005/06 (Sport England, 2009).

Mountain biking

The evolution of the mountain bike has allowed cyclists to leave the road and enter the realm of NCAs, adding a distinctive self-propelled way of enjoying the challenges of natural land features. Throughout the twentieth century various modifications were made to the standard 'safety bicycle' in order to equip it to handle rough terrain for a variety of purposes. The breakthrough for the mountain bike is traced to developments in California in the late 1970s and early 1980s, leading to the production of 'the single most significant development in the history of the bicycle in the last 100 years' (Eassom, 2003, p. 194). In the last two decades mountain biking has moved into the mainstream, and the latest high-tech, light-weight machines sport suspension systems, shock-absorbing wheels, wide knobbly tyres, strong frames and gears which equip them to withstand the rigours of being ridden just about anywhere: up steep slopes and down again; through mud, streams and sand; over rocks and logs; and along narrow winding tracks.

Minimally modified areas such as fire roads, animal and hiking paths are favoured for mountain biking, with narrow and winding 'singletrack' in forested areas being particularly popular for the challenges it poses in attractive natural surroundings. The mountain biker must make constant decisions regarding steering, braking, balance and gearing in order to avoid obstacles and stay on the trail. The techniques for climbing steep slopes include walking (also known as 'first gear'); spinning, in which the rider stays in a sitting position and pedals quickly in a low gear; and honking, where the rider stays in a higher gear and stands up on the pedals so as to apply the maximum amount of force on the pedals. Technical skill in manipulating the bike and shifting body weight is also required to negotiate tight corners, descend ledges and large rocks, to 'bunny-hop' over obstacles and to ford streams.

But while the mountain biker might be absorbed in the execution of 'gnarly' moves, US mountain biker Lee Bridgers (2003) argues that the 'single most important draw of riding a mountain bike is NATURE' (p.186). Without this fundamental focus overriding the seduction of technical virtuosity for its own sake (and, perhaps, for the sake of impressing others) Bridgers believes that a lack of respect can lead to the degradation of the natural environment and physical harm to the participant.

New Zealand mountain biker Damon Trenwith enthuses about a multi-day ride with its combination of exhilarating physical challenge and natural beauty.

> It's probably our country's longest piece of continuous singletrack. Seventy-one kilometers of mountainous, one-way mountain bike track. A trail that incorporates grueling climbs, breathtaking views and fast, sweeping downhills that run along ridgelines and hillsides sounded good. Couple that with trails that weave through lush forests and descend into secluded coastal bays with loads of fresh air.

It all equates to the Queen Charlotte Track, an absolute must-do for any keen cyclist – and irresistible to myself.
(http://www.dirtworld.com/features/TrailStory.asp?id=592, retrieved 16 January 2009)

The first world mountain biking championships took place in 1987, and these competitions are now recognized by the International Cycling Union and include cross-country, downhill, four-cross or mountain-cross (with four bikers racing downhill on a prepared track) and trials events. In the United States, 7.1 million people went mountain biking in 2009 (Outdoor Foundation, 2009a, 2010).

Orienteering

Orienteerers find their way across varied terrain, usually on foot, using a topographical map and a compass to follow a preset course consisting of 'control points' plotted on the map. While orienteering may be undertaken alone or in groups for the simple satisfaction of successfully navigating through a natural setting, participants often compete with each other to complete courses in the shortest possible time. Orienteering was developed in Scandinavia in the late nineteenth century, originally as a military exercise. Over the course of the twentieth century, as its popularity has grown and spread, it has become a highly organized sport with national and international championships and a number of related offshoots including ski, canoe, scuba and mountain bike orienteering, and rogaining (Rugged Outdoor Group Activity Involving Navigation and Endurance) where teams compete on courses for up to 24 hours at a stretch. Trail orienteering places navigational accuracy above speed and thereby allows participants with limited mobility to compete on equal terms.

Orienteering courses may be set in various locations including urban parks, but as an NCA the hobby is typically pursued in forested areas and open hill country. Whatever the nature of the terrain, orienteering adds another dimension to the natural challenges involved in hiking and trail running; that is, the execution of the activity with the additional element of quick, accurate navigational problem solving. According to Loland & Sandberg (1995), the skills required for orienteering are both physical and intellectual. In addition to the endurance, strength and technique that running demands, orienteerers must be develop route-choice, navigational and psychological skills.

Having good route-choice skills means that one is good at individually selecting the best route from studying the map before physically setting out on it. A runner with good navigational skills is good at sticking to the selected route by individual navigation with help of the two approved tools; map and compass. Of these two, map-reading skills are considered the most important. The psychological skills are similar to what is required in competitive sport in general and consist among other things of being able to stand the pressure of the competition situation and of keeping the concentration during the whole race.

(Loland & Sandberg, 1995, p. 232)

Aside from the thrill of competition, studies show that the natural beauty of orienteering locations, and the sense of adventure and discovery that such locations facilitate, are strong attractions and sources of satisfaction for participants (Koukouris, 2005). Orienteering is also seen as an enjoyable means of developing the necessary navigational skills for a range of other NCAs including hiking, mountaineering and caving. In Australia, 124,000 people went orienteering in 2008 (Australian Sports Commission, 2008), while monthly participation in the United Kingdom was 9300 in 2008/09 (Sport England, 2009).

Geocaching

Geocaching has been described variously as '21st-century hide-and-seek' (Schlatter & Hurd, 2005) and 'high-tech treasure hunting' (http://www.geocaching.com, retrieved 14 December 2008). Like orienteering, it offers participants a navigational challenge in a natural setting. In the case of geocaching, a GPS device is used to locate a 'cache' which has been hidden by another geocacher. Clues to the location of caches, including the relevant satellite coordinates, can be found online. Details about finds and stories of geocaching adventures are recorded on the same websites. A cache in its most basic form will consist of a container with a log book (for recording the find) inside. More elaborate caches may contain small objects which can be swapped with similar objects carried by the geocacher, or 'travel bugs' which are items intended to be moved from cache to cache.

Geocaching is a very recent NCA derivative. It was 'created' in 2000 by GPS enthusiasts exploiting the substantially improved accuracy of satellite technology for recreational purposes. In the space of a few years, geocaching has spread throughout the world, from North America to Europe, Australasia, South America and Africa. Caches have even been placed in Antarctica. According to the main website on which caches are recorded – www.geocaching.com – there are more than 1.4 million geocaches worldwide, which have been found by over four million people (http://www.geocaching.com/articles/Brochures/footer/FactSheet_Geocaching.pdf, retrieved 7 July 2010).

Geocachers must have the ability to use the internet and a GPS receiver, as well as read a map. Depending on the chosen cache, participants must also be able to negotiate the terrain in which it is hidden. There are caches to be found in urban areas and close to roads. However, where geocaching is to be enjoyed as an NCA, the geocacher will select a cache according to their preference and skill level, which could involve anything from a short hike or mountain bike to more remote locations requiring scrambling, glacier travel and other mountaineering skills. Caches have been placed in the desert, on islands and underwater. An indication of the nature of the challenge involved in finding a particular cache will be given on the website entry by a difficulty and terrain rating. Once the GPS coordinates have been reached, observational skills are required to actually find the hiding place of the geocache, which is not buried, but could be under a pile of sticks or a rock, or up a tree. There is also the element of inventiveness and creativity for those who hide caches.

Geocaching may be undertaken independently by an individual or group as a way of adding another layer of challenge and exploration to the core activity of an NCA. Some are inspired by the challenge of amassing as many finds as possible, or by finding challenging caches that few other people have managed to locate.

Jim Brakefield from Clay, Alabama, explains how he and his wife became enthralled by geocaching.

> It was so exciting to take that little bit of coordinate information, program it into the GPS and let it tell us where to go. Of course, the hunting techniques evolved after finding a few more caches, but still the simplicity of the GPS and the fun of the hunt is what lured us in.
>
> Geocaching has taken us out of the house and to some of the most beautiful places in and around the state of Alabama . . . .
>
> The best part is that geocaching takes us off the beaten path to explore more thoroughly the hidden sights to be seen and enjoyed.
>
> Not only has geocaching gotten us out of the house, we are getting some much needed exercise. We are climbing up and down hills, jumping creeks, hiking the backwoods and loving every minute (well, mostly every minute!) of it. On some of our more strenuous cache hunts we come home exhausted but always fulfilled!

Another 'best part' of geocaching is meeting fellow geocachers with a kindred spirit and appreciation of the outdoors....

Geocaching has opened up a whole new world for us to enjoy. We love the thrill of the hunt (and more especially, the find!), the places we are going, the sights we are seeing and the people we are meeting! (http://onlinegeocacher.com/articles/What_Geocaching_Has_ Done_For_Us_04,21,08,12,04,21, retrieved 15 January 2009)

Clubs and associations have begun holding competitive geocaching events, similar to orienteering. Caches may be hidden in a defined area and participants compete, either individually or in teams, to find as many caches as possible within a set time period, with a point value assigned to caches depending on their difficulty rating and distance from the starting point. In another version, teams race to complete a course with a set number of caches as waypoints.

Power kiting and sandyachting

Land-based NCAs that simultaneously harness the power of the wind are mentioned here only briefly due to space limitations and because closely related – and more popular – activities are discussed in detail elsewhere in this book (see kitesurfing and sailing in Chapter 3, and snow and ice kiting in Chapter 6).

The same kites used for traction on water, snow and ice may be used on dry land for the activities of skidding or skudding (being dragged by the power kite on one's feet or back), kite jumping (also known as moon-walking or 'getting airs'), and buggying or karting. The beach is an ideal setting for these activities, or in the case of skidding, wet grass. Alternatively, for kite buggying, the open spaces and smooth surface offered by desert or salt flats make for a good location, particularly if they are accompanied by steady winds (Boyce, 2004). Kite buggies are typically single seat, three-wheeled vehicles, in which a driver sits, holding the kite in his or her hands while steering the single front wheel with the feet. In the United Kingdom in 2005/06, 6100 people were power kiting on a monthly basis (Sport England, 2009).

Hard, sandy beaches and dry salt lakes are also ideal for sandyachting, a subset of the broader activity of landyachting which may also take place on artificial surfaces such as airfields or parking lots, taking it outside the realm of NCAs. Sandyachts have the same basic configuration as kite buggies, with the addition of a sailing rig, and can be sailed at three or four times the wind speed (Parr, 1996). Competitions are overseen by the International Land and Sandyachting Federation. Parr (1996, p. 868) estimates that in the mid-1990s there were about 150 active sandyachters in the United Kingdom, and almost 1000 in the United States.

Wilderness horseback riding

While there are a number of different styles and forms of horseback riding for sporting or recreational purposes, it is pursued as an NCA in the form of wilderness or trail riding. Trail rides, which take the horseback rider out of the paddock or arena and into the backcountry to follow forest, mountain or desert trails, old logging roads or bridle paths, are undertaken for both pleasure and competition. Like hiking, horseback riding may be combined with a number of other NCAs including camping, hunting, fishing and orienteering. A special element of trail riding is the bond that develops between horse and rider, and horses – not unlike humans – need to be conditioned by a programme of exercise, feeding, medical and general care. A good trail horse should be a 'powerful, courageous horse with great lung capacity, heart, and endurance' and have an even temperament (Westbrook & Westbrook, 1966).

Throughout history, horses have been ridden across country as a matter of necessity. In the United States, for example, the tradition of long-distance backcountry horse riding is recognized as part of pioneering history. With the rise of motorized transport and the corresponding decline in the use of the horses for transportation, people have continued to ride horses for pleasure and horseback riding has become established as a hobby.

In backcountry horse riding the nature challenge is met by both horse and rider together. This requires of riders not only an appreciation of the terrain – whether it be rocky, sandy, boggy, or involve stream crossings – and basic backcountry skills common to other land-based NCAs such as navigation and reading weather signs, but also that they develop a level of understanding of and a bond with their horse. Long rides demand endurance from both horse and rider. Western riding style, which is easy to maintain for both horse and rider over extended periods, is recommended (Westbrook & Westbrook, 1966, p. 53). The rider must also understand the horse's condition, be able to set a satisfactory pace, and watch out for signs that the horse may be tiring. Varying the horse's gait helps to increase its endurance by resting certain muscles and allowing the horse to catch its breath with intermittent periods of slow trotting or jogging. The rider must be alert to obstacles and potential hazards such as stones and holes and the hooves of other horses when riding in groups.

Competitive events include endurance races, competitive trail rides (CTR) and judged trail rides, in which horses and riders can cover substantial distances of varied terrain, with altitude and weather conditions contributing to the challenge. In all events, veterinarians strictly monitor the horses' condition, checking their fitness, respiration and heart rates at regular intervals, before, during and after a ride. Horses are also checked for signs of fatigue, sickness, dehydration and lameness, and any deemed not 'fit to continue' are withdrawn from the event. In endurance races, which can cover distances of 40–160 kms in one day, the winner is the first horse and rider team to complete the course and still be 'fit to continue', while a special award is given to the horse that finishes in the 'best condition'. CTRs vary from one to three days in length, covering between 15 and 70 kms a day. The ride may be timed, in which case competitors must complete the course within a optimum time period, with points being deducted for being either too fast or too slow. Competitors are also evaluated according to various criteria including, for example, the horse's heart rate at the completion of the ride, the rider's horse management and the skill by which obstacles are negotiated. In a judged trail ride, competitors are evaluated at various points along a trail according to their success in negotiating obstacles, and speed is not a factor.

Other variations on these events include multi-day pioneer rides over historic trails and Ride and Tie, which combines trail running and endurance riding with teams comprised of a horse and two riders/runners who take turns riding while the other runs over a set course that can be up to 160 kms long.

Sixteen-year-old Katie DeVoe, an experienced endurance rider, describes her first Ride and Tie event.

> ... I had a total blast. There is nothing more exhilarating than running so fast downhill that the horse cannot catch up to you. I know that I will never forget sprinting across the finish line with Fancy cantering behind me. Nor will I forget all that I learned. Running on the same trail as my horse gave me an all new perspective of what my horse needs to do to maintain balance and speed. I now understand why horses do not like to trot on slanted trails, or trot up hills when they are tired. I know just when the hill is too steep for my horse to

safely and comfortably trot down it. Most importantly, I learned that three individuals can be so focused in working towards a common goal that the miles will just slide on by.

(http://www.rideandtie.org/followinghoofprints.pdf,
retrieved 16 December 2008)

Off-road motorcycling and 4WD

Off-roading is the familiar term for motorized land-based NCAs, where specialized vehicles are used to traverse natural terrain such as sand, gravel, riverbeds, mud and rocks. Off-roading, therefore, is the NCA offshoot of recreational 4 Wheel Drive (4WD) drivers, and All-Terrain-Vehicles (ATVs) and motorcycle riders. While many other land-based NCA enthusiasts prefer the physical challenge and natural quiet of self-propelled activities, an advantage of off-roading is that it is accessible to participants from a broad range of age groups and physical abilities. Off-roaders, however, will never be able to go as far into the backcountry as those on two legs, as they are generally confined to suitable tracks, river beds, beaches and similar open country. In addition, many natural areas are closed to off-roading because of the noise level, erosion and danger that it poses to self-propelled alternatives.

Vehicles designed for the specific purpose of off-roading generally have a high ground clearance, rugged tyres and a low gear ratio for tackling steep hills or crossing rough terrain. 4WDs may also have a locking front and rear differential, which can increase traction. The development of the Land Rover in the 1950s and 1960s, and the subsequent emergence of cheaper Japanese alternatives, boosted the popularity of recreational off-roading in 4WDs. Early off-road motorcycles were built for motocross competitions in the 1920s, while ATVs, which are ridden astride like a motorcycle but have three or more wheels for additional stability at slow speeds, appeared in the 1970s.

Like horseback riding, the essence of off-roading is overcoming a natural challenge through a successful partnership: in this case between driver/rider and machine. The potential (and limitations) of the machine must be clearly understood and utilized (or overcome) when climbing, descending or traversing steep slopes, fording streams, tackling mud bogs and soft sand.

Cockroft (1997, p. 47) gives the following advice about the various kinds of surface that off-roaders must negotiate.

Muds vary from thin slimy coatings to deep wallowing pools. The texture varies dramatically according to the local soil structure from which it is formed and the amount of water it contains.

Likewise beaches, river sands and dune formations all have their unique properties. Fine dry sand may make for difficult if not impossible driving but it will seldom prove dangerous. On the other hand, course wet sand can swallow a vehicle whole and then some.

River pebbles vary according to size and underlying strata. Large boulders will provide firm if uncomfortable traction. As the size diminishes, instead of riding over the top, wheels may begin to dig down and the vehicle may become hopelessly bogged down.

In many instances even the most experienced driver will find it impossible to judge a surface simply by looking at it. Experience will give you the ability to react to the 'feel' of the track; how you react can be very important.

An ability to read surface conditions is important for making the correct decisions about the optimal route to take, which gear or ratio to use and how much throttle to apply in order to maintain momentum without causing the wheels to spin. Off-roading also requires specific braking techniques, as water or mud may render brakes ineffective, while normal braking on steep slopes or slippery ground can cause a vehicle to slide.

Off-road motorcycling, in addition, requires good physical fitness, as explained in a motorcycling manual:

On the trail you will constantly stand on the bike's footpegs. Your arms and legs will manhandle the bike and prevent the shock of the trail from reaching your body and brain. You must often lift the front tire on the approach to a hazard, toss or bounce the bike left or right around a hazard, and hang on for dear life or lose control of the bike.

(Bennett, 1995)

The four most common types of trail bikes are enduro, motocross, trials and cross-country, according to the type of racing competition for which they were designed. In motocross the goal is to complete a prepared course in the fastest time. Trials riding tests control rather than speed, as competitors negotiate various obstacles on a course and receive penalty points for putting a foot down, stopping and/or walking the bike, or riding outside the set course. Cross-country bikes are designed for long-distance races such as the famous Paris-Dakar and California Baja 1000. Enduro are off-road time keeping events, whereby competitors attempt to complete various stages of a course within a pre-defined schedule, with points being deducted for being either too early or too late. Other events include beach and desert racing, which are contested by a variety of off-road vehicles.

In a study of ATV enthusiasts in the United States, Mann & Leahy (2009) found that riders enjoyed the natural setting of their activity in much the same way as non-motorized recreationists. In particular, they felt a connection with nature through close encounters with wildlife, spectacular scenery and the smells of the woods that they rode through. Additionally, ATV riding provided access to remote areas for hunting and fishing, and was an accessible activity for people with health conditions that prevented them from participating in other outdoor activities. Club events, including working on trails and other community service, provided an agreeable social dimension to this pursuit (M. J. Mann & Leahy, 2009).

Conclusions

Land is perhaps the broadest and most accessible of our NCA categories, geographically, economically and aesthetically. Although they are essentially encountering the same natural elements, mountaineers, cavers and desert racers are drawn to activity settings with their own very distinctive appeal. While caves, canyons and mountains may not be easily accessible in all parts of the world, most in the developed world live within striking distance of an area amenable to hiking. From the standpoint of consumption, although horseback riding, off-roading and mountain biking require some investment in terms of equipment, the hiker, scrambler or mountain runner needs little more than appropriate footwear to participate at the most basic level. A simple, relatively unmediated interaction with nature may indeed be what is sought by the NCA enthusiast who heads into the backcountry with no more than a backpack and a few essential supplies. On the other hand, those seeking a nature challenge with a greater technological dimension have ample opportunity within the land-based NCA spectrum. While the more technical pursuits such as climbing and caving require the purchase of additional equipment, the cost need not be prohibitive. Second-hand gear is often available and clubs frequently provide the means to borrow or hire expensive items such as climbing ropes. Meanwhile, dedicated participants often pride themselves on 'making do' with what they have rather than purchasing the latest gadgetry. Such innovation can be a valued skill in itself as it demonstrates hard-won experience and self-reliance.

The social dimensions of the NCAs covered in this chapter vary widely as well. While the most physically demanding and technically challenging are dominated by young, male participants, activities like hiking, orienteering and geocaching are enjoyed by cross-generational groups from a wide variety of backgrounds. Indeed, family hiking trips are often the entry-level activity which leads to a life-long passion for land-based NCAs.

Considering their sustainability, whether it is feet, wheels or hooves, all modes of traversing the land can cause varying degrees of damage to vegetation, as well as soil erosion and compaction. Motorized vehicles, in particular, can adversely damage and pollute the environments in which they are used, as well as being detrimental to the social and cultural values of other land users (Randall et al., 2006). Further negative impacts may be caused by litter and human waste being left in remote areas, as well as mountaineers leaving equipment such as pitons, bolts and ropes in place to facilitate future ascents or descents. This said, most groups of land-based NCA enthusiasts aspire to protect the landscapes they enjoy and promote the backcountry ethic of minimizing their impact on the environment by avoiding disturbance to flora and fauna, carrying out rubbish and responsibly disposing of human waste. A good example is the geocaching community's Cache In Trash Out (CITO) ethos and initiatives (see www.geocaching.com/cito).

5
Flora and Fauna

The plants, trees, animals, birds and other living things on Earth, unlike the ice and snow discussed in the next chapter, generally have a good reputation. True, when in the wilds (read all of nature), it is wise to worry about poison ivy, venomous snakes, disease-carrying mosquitoes and other threats. But we tend not to hate them, as some of us do the cold that makes for ice and snow. Rather our attitude towards harmful flora and fauna is typically one of caution. Know what harm you may encounter in nature, and protect yourself from it by relying on plant recognition, high-topped boots and insect repellent, respectively.

Some people, almost all of them city dwellers, learn to fear particular features of nature they believe they may meet in the wilds. Certain large mammals and reptiles of various sizes commonly head up their lists of its dangers. 'What if I should encounter a shark, crocodile, grizzly bear, or wild boar'? Experienced and knowledgeable people pursuing leisure activities in the habitats of such creatures respect them – they use caution – but they do not fear them (true, they would be fearful should they, despite precautions, meet one face-to-face).

This chapter concentrates on the NCAs involving particular flora and fauna as observed, used or studied in water, on land and in the air. Participants in these NCAs are cautious when need to be, but certainly not fearful. Their dominant attitude is positive; they are oftentimes in awe of that part of nature on which their hobby is based.

The following hobbies and amateur pursuits are covered in this chapter. Nature poses a distinctive challenge in each:

- **Flora:** nature photography and painting, gathering and studying mushrooms, collecting and gathering natural objects, and amateur botany.

- **Fauna:** fishing, hunting, bird-watching/ amateur ornithology and collecting insects/amateur entomology.

Not covered in this chapter are a number of kindred casual leisure activities centred on certain flora and fauna, some of which are most accurately classified as kinds of sightseeing. Examples include commercial tours to watch whales, polar bears and sea birds as well as personal outings to observe for pleasure wild flowers, salmon runs, bird migrations and sea shells. Also excluded is the casual leisure activity of gathering wild edibles, among them, raspberries, blueberries and strawberries.

Flora

Most of the hobbies and amateur activities examined in this section require removal of either part of a plant or the entire plant. Not so, however, with nature photography and painting.

Nature photography and painting

This category of NCA is treated of here, because most photography and painting is of flora rather than fauna. But, since some amateurs go in for animals (birds, fish, reptiles etc.) as subjects, they too will be discussed here. Most of what is considered in this section deals with productions of still life. Nonetheless filming and videotaping moving fauna are also possible at times.

Technically speaking, nature photography dates to the invention of the camera, since the early devices were able, with sufficient light, to record both natural and artificial still life. But it was not until late in the nineteenth century that photography evolved to a point where it was simple and efficient enough to be an accessory for serious leisure activities. For this to happen, cameras had to be portable and equipped with the magnesium flash. Portability obviated the need for weighty tripods. Moreover, though long exposures were still necessary, it had become possible to record as a non-moving image a subject in motion.

Nature photography has numerous branches commonly known as: aerial, landscape, seascape, cloudscape, wildlife, underwater and astrophotography. Taking pictures of plant life (trees, plants, flowers and bushes) tends to be regarded, more vaguely, simply as part of nature photography. Today professionals and amateurs pursue their devotee work and serious leisure interests in all these branches, although not always in equal proportions. Thus, aerial photography, because of the costs of getting airborne, is mostly a professional interest expressed in photographing industrial, military, archaeological and community sites or in providing photographic images for map making.[1] Meanwhile, the comparative ease of getting into the outdoors to photograph or film plants and animals is among the reasons why it attracts more amateurs than professionals. Astrophotography, a lively field for amateurs, is one of the activities that had to be omitted for space reasons.

What makes nature photography, film craft and videography serious leisure or devotee work? The extensive skill, knowledge and experience which underlie these pursuits is found, in the main, in using the camera (e.g., assessing lighting, film speed, background of subject), knowing the relevant habits of the subject, applying aesthetic principles, and managing the meteorological conditions, geographic terrain and clothing requirements encountered in and needed for reaching the subject. The modern digital single-lens reflex (DSLR) camera has simplified camera use, though primarily for novices in this field, for the ordinary single-lens reflex (SLR) camera is still preferred by advanced enthusiasts.

Thus there is a career of personal improvement along these lines that comes with experience in each specialty. Perseverance is frequently required in, for example, getting to remote places where wild flowers grow, finding the best specimens once there and waiting for optimal lighting in which to photograph them. The distinct identities of nature photographers, their social worlds and the rewards of their core activities are brought out below in the sections about each type.

Plant life

The core activities of plant life photography are, first, setting up and taking pictures and, then, creating digital or paper productions of the subjects photographed. But the general activity of this kind of photography includes far more than this. For instance, these photographers must get to and from the places where they find their floral subjects, which might be by foot, automobile, bicycle or boat, to mention just some of the modes of transportation. They must also buy supplies from time to time and occasionally buy a camera. Since some belong to clubs and some exhibit their photos, the general activity of nature photography includes for them participation in exhibitions, competitions and club meetings and events. Moreover many photographers read books and magazines related to their art, which some consider a core activity but others consider part of the general activity. Finally some take commercial nature photography tours, while others with the means organize personal expeditions to faraway places with the principal goal of photographing new subjects.

The complexity of the core activity of nature photography is evident in the following description of taking a picture of a white water lily and processing it:

> I used a tripod for added stability but I discarded the tripod after this shot, it was simply too difficult. In the lightroom, I adjusted the landscape mode setting and the white balance. I adjusted the light levels so that the flower was brighter against a dark background. The pond was under tree cover but sunlight was dappled, in places bright and in others filtered. I cropped and tilted the photo to make up for the tough angle I had to use.
>
> (source: http://burstmode.wordpress.com,
> 27 September 2009, retrieved 13 October 2009)

People in the West living in an urban area and enamoured of photographing plants or trees (or both) have available one or more clubs that support this interest. These groups are, however, devoted to nature photography, in general. The North American Nature Photography Association (NANPA) claims to be the first and premier group of this sort in North America that is committed solely to serving the field of nature photography. It has amateur, professional and corporate members (2400 in 2001), who together, pursue the following mission:

- NANPA promotes the art and science of nature photography as a medium of communication, nature appreciation and environmental protection.
- NANPA provides information, education, inspiration and opportunity for all persons interested in nature photography.
- NANPA fosters excellence and ethical conduct in all aspects of our endeavours.

(Source: http://www.nanpa.org,
retrieved 13 October 2009)

A similar local proliferation is evident in Europe (usually by country or city), where the overarching association is the United Kingdom and Europe Nature Photographers (UKNP). Its mission is similar to that of NANPA. The Australasiaforum.net organizes nature photographers in that region primarily by means of online activities and information. The Australasian region is defined as Australia, New Zealand, Papua

New Guinea and surrounding Islands and for purposes of the forum also includes Antarctica, Indonesia and the Philippines. Its forums were established to provide the Australasian area of the world with a place to discuss and post images relating to nature photography. All these associations offer some kind of reading material (e.g., journals, articles, references to books), advice on cameras and photography and online links to useful websites.

The awe-filled experience of photographing wildflowers is captured in the following testimony by the photographer (procedure described above) who had taken a picture of a white water lily in the wilds:

> These shots were simply not easy to get. A considerable amount of reaching, stumbling and balancing had to be done. Was it worth it? You bet. These things are gorgeous. They are loud flowers and remind me of my childhood. Growing up in Trinidad, the rhythm of life was loud and colorful. The air was always full of laughter. Looking at these beautiful flowers, how could you not feel anything but good? And if it cost me a little sweat and a slight backache... c'est la vie.

Another photographer described a hike he had taken and the role of photography within it:

> The hike follows the Delaware Canal Towpath from Upper Black Eddy, PA to Tinicum County Park and back. Along that stretch of towpath, you will find a covered bridge, some great birding, butterflies, marshy areas loaded with frogs that play a one-note symphony for hours on end and some beautiful wildflowers. Suddenly I remembered the Daylily images that I made along the towpath last summer and had to go digging through my back files in search of them.... [They were] a memory of one of my favorite local hikes.
>
> (source: http://itsmynature.wordpress.com/category/
> daylily, retrieved 13 October 2009)

Landscapes

Landscape photography centres on the natural features of the earth, including its trees, mountains, prairies, waterfalls and water courses. Wikipedia says that 'many landscape photographers show little or no human activity in their photos, striving to attain pure, unsullied landscapes that are devoid of human influence, using instead subjects such as strongly defined landforms, weather, and ambient light' (http://en.wikipedia.org/wiki/Landscape_photography, retrieved 14 October 2009). Both amateurs and professionals go in for landscape photography, some of whom define their field differently than we have here.

There is plenty to read for both types. Amazon.com lists hundreds of books on the subject, many of them manuals on how to be a better photographer in this field. Among the periodicals are *Outdoor Photographer* and more general ones such as *American Photo, Shutterbug* and *Practical Photography Magazine*. Moreover by joining a photographic organization members often receive a journal. Thus, the Photographic Society of America (PSA), with over 5000 individual members (and over 400 camera club members), says:

> enjoy your hobby with members in 60 countries throughout the world. Established in 1934, it is the largest association of its kind, bringing together professional and amateur enthusiasts of all ages and levels of achievement. As a non-profit organization, its mission is to promote and enhance the art and science of photography in all its phases, among members and non-members alike.
>
> (http://www.psa-photo.org/aboutPageDisplay.asp?DivID=8& menuID=1&pageID=56, retrieved 14 October 2009)

The PSA also provides members with its monthly, the *PSA Journal*. The Royal Photographic Society (Britain, 9657 members in 2005), the Società Italiana di Caccia Fotografica (Italy, no membership data available) and the Australian Photographic Society (2100 members in 2008), which offers two periodicals, *Image* (bimonthly) and the *Australian Photography* (monthly), number among the many organizations across the world established for amateur and professional photographers.

The organizations just described are for all nature photographers. So landscape photographers interested in joining a club or other organization must seek out either one of them or a general camera club. Organizations devoted exclusively to landscape photography appear to be rare. Still the Internet is also a main resource for these hobbyists. For example, the Photography Webrings (http://www.photography-webrings.net) have been established for landscape, nature, travel, fine art and digital photography, among others. They bring interested visitors to personal photo sites, enabling them to become a part of a cyber-community of photographers sharing similar interests. They also provide a convenient way to find photo sites having similar themes and explore new genres of photography visitors may not have yet discovered. Also of interest is the general *Nature Photographers Online Magazine*, sponsored by the Nature Photographers Network (http://www.naturephotographers.net/members.html). It offers articles, galleries and forums.

Several characteristics of landscape photography as serious leisure are evident in Tsubakuro's experiences in Japan:

> For mountain photography, time is of the essence. One studies the maps in order to plan where to be at the right time for the right light and how to get there by the swiftest and least taxing means. The mountain photographer requires time to reach his goal, survey his chosen site, set up his equipment and be ready for the moment. At least this holds true when we are talking about sunrise and sunset photography in the mountains. The rest of the day (and night if one chooses) can be spent at a little more ease, exploring the nature, seeking new views to capture. In a day, the photographer will unlikely walk too far, choosing instead to spend his time searching for subjects and scenes fit for his lenses.

> For climbing, time is a very important factor. The climber studies the maps to check the routes and times, planning how much time will be required to reach check points. The climber wants to move swiftly and lightly. For a day of climbing much energy will be spent on moving up mountains and crossing ridges. Being weighed down by a heavy pack is undesirable. A light pack with food and drink and a light-weight camera are enough. Use the daylight hours efficiently to reach the summits and get back in a reasonable time to enjoy dinner at your tent.

> I used to be only a photographer. Reaching the summits was not a concern for me. But the last few years have been different. Reaching the summit has become almost as important as bringing home a good collection of photographs. That is one reason why I shoot less these days. Another reason is that I shoot 35mm, medium, and large format on most outings. That means a heavier camera pack and more time spent switching cameras and lenses. Especially using a 4×5 camera takes time. Thus my daylight hours have to be carefully balanced between moving up and over summits and being in the right places at the right times to capture the scenes I desire to bring home. Often to my frustration, the time pressure of reaching peaks and getting photographs often leaves me in a time pinch.

> (http://tsubakuro.wordpress.com,
> retrieved 24 October 2009)

Wildlife

The subjects of wildlife photography and motion pictures include birds, fish, reptiles, mammals and insects. Wildlife photographers interested in meeting kindred spirits, finding advice, reading on their activity and the like must search for these within the websites, literature and organizations of generalized nature photography. We could find no magazines devoted exclusively to this hobby, apart from those of the National Wildlife Federation, an American group, which publishes four periodicals for children. Nevertheless this genre of photography constitutes a hobby of its own. Christina Craft, who runs an online stock photo library, comments on the detailed preparation needed for wildlife photography and the wonderment that comes with doing it:

> Nature and wildlife photography are different from other genres. A one-second different in light can change a scene from mundane to wow! My passion for wildlife photography stems from a love of animals and nature. I believe that passion for your subject is very important. That's why I've devoted every trip over the last five years to wildlife. I study each animal and find out everything about its behavior and habitat. Then I seek the animal out, while not intruding on its natural path. I love making connections with wildlife. There's nothing like having a grizzly bear stare straight into your eyes from only a few feet away or to hear the giggles of tiny squirrel monkeys as they tickle each other on a palm tree while you're clicking the shutter.
>
> (http://photography.naturestocklibrary.com/nature-wildlife-photography.html, retrieved 14 October 2009)

Underwater

Photographing or videotaping subjects (e.g., fish, shipwrecks, vegetation, land formations) in lakes, rivers and the sea is commonly done while scuba diving, though swimming and snorkelling also offer limited opportunities for this activity. On the Seafriends website, underwater photography page (http://www.seafriends.org.nz/phgraph), it is noted that diving helmets had to be invented before such photography could properly come into its own. So it was not until 1856 that William Thompson, using a primitive underwater camera, could take photos of seaweed near Weymouth, England. The history page of this website contains a chronology from this date to the present of major advancements in underwater photography, including its military uses and the famous oceanographic work of France's Jacques-Yves Cousteau. Also available here is a wealth of practical information and useful links. In addition see Edge Underwater Photography at http://www.edgeunderwater photography.com (retrieved 14 October 2009).

As with the other kinds of nature photography, underwater photography – both still and motion – requires more knowledge than how to select a suitable camera (a waterproof housing being an obvious feature) and how to use it effectively under water. These hobbyists must be competent scuba divers as well. Furthermore they must be able to work with a buddy (a requirement in scuba), operate under a variety of meteorological conditions and their effects (e.g., clouds, waves) and manoeuvre in and take advantage of currents and tides.

In this area, too, the literature is relatively thin. There is *Underwater Photography* (http://www.uwpmag.co.uk, retrieved 14 October 2009), a free, bimonthly, web-based periodical devoted to digital equipment news and reviews, location reports, personality profiles and portfolios, tips for better pictures and classified advertisements. Some specialized clubs and associations exist as well. The Northern California Underwater Photographic Society (NCUPS) (http://www. ncups.org/links.html, retrieved 14 October 2009) lists on its Links page 13 other such societies. Most are in the United States, with one each located in Canada, Australia and New Zealand. Participants living in places without such organizations but having organizations devoted to general scuba diving can usually find some similarly minded enthusiasts here (scuba is discussed in Chapter 3). The NCUPS Links page is voluminous; it also offers sections on books, periodicals (all more general than underwater photography), resources, education, conservation sites and marine research, to mention only some.

Cloudscape

This kind of nature photography and motion picture-making focuses on clouds of different shades and colours, in particular, and on the sky, in general. Having a clear view of the sky is essential in this activity, which sometimes favours black and white photography. According to the website Cloudscape Photography it is most commonly pursued in Norway, Cuba, Sardinia, Vietnam and the counties of northern Germany. The contributions of famous twentieth-century cloudscape photographers like Ralph Steiner and Robert Davis have brought immense acclaim to this photographic genre (http://www.mapsofworld.com/referrals/ photos/forms-of-photography/cloudscape-photography.html, retrieved

14 October 2009). Léonard Misonne (1870–1943), a Belgian, was one of the earliest cloudscape photographers, noted especially for his black and white photographs of heavy skies and dark clouds.

These photographers may keep in touch with each other through Cloudscapes, a photoblog (http://cloudscapes.antville.org, retrieved 14 October 2009). Otherwise, for company and information, they along with the wildlife photographers must join either a nature photography organization or a general photography club. They must also turn to the nature and general photography periodicals to read the occasional article on their specialty.

Nature painting

Artistic painting and drawing done at amateur and professional levels are highly skilled pursuits based on considerable specialized knowledge of, among other things, the medium (e.g., oil, ink, charcoal, watercolour), brushes and pencils and surfaces (e.g., canvas, wood, paper) artists use to portray creatively their ideas. Conceiving of a paintable subject, making a sketch of the painting to result and completing that work from the sketch are core activities in this pursuit. As with landscape photographers, nature painters must know the relevant habits of their subjects, apply aesthetic principles, and manage the climatic conditions, geographic terrain and suitable clothing encountered in and needed for reaching them. Unless the painter is working from a photograph of, say, a bird or animal, that subject is usually inanimate – a flower, tree or broad paysage. When wildlife painters work from their own photos, they must also be wildlife photographers, conducting their fieldwork much as these amateurs and professionals do.

The website Ethnic Paintings (http://www.ethnicpaintings.com/painting-trivia/landscape-painting2.html, retrieved 17 October 2009) states that landscapes can be observed in the remains of the frescoes of first-century Pompeii and Herculaneum. However, it was not until the fifteenth century in Europe that landscape painting became a recognized art form, usually with a religious subject. Joseph M.W. Turner, an early nineteenth-century English artist, is credited for laying the foundations of impressionism. Meanwhile American landscape paintings came to dominate the art scene of the 1820s and beyond. The unexplored and unexploited wilderness in North America held allure for painters everywhere. Some of them belonged to the loosely organized Hudson River school, whose members painted landscapes between 1825 and 1875. Cloudscape, as an artistic interest within the landscape type, has been around for centuries. It came in to its own in the early nineteenth century through the works of Turner, countryman John Constable and Caspar David Friedrich, a German artist (Gedzelman, 1989).

Today drawing and painting clubs abound in nearly every Western city. And most of them are now equipped with websites where painters of various specialties, landscape and nature included, may meet fellow enthusiasts, show their works, learn tips about their art, receive news about competitions and exhibitions and possibly receive a newsletter. Furthermore there are available in these cities and in the countries where they live numerous magazines and books for the serious painter.

There are no websites specifically designed for nature painting. Instead painters of this genre must go to the general artistic painting sites where their specialty is, however, well represented. One established general site is Wet Canvas (WC). It claims to be the largest community on the Internet for visual artists. It was founded in 1998

> in an effort to better leverage technology to assist visual artists in sharing information and making new contacts and friends.... Our online community is comprised of artists of all levels, ranging from Sunday painters to artists who exhibit in some of the finest galleries in the world. Whether your interests lie in learning new techniques, experimenting with alternative or cross-over mediums, gaining critiques from fellow artists, or getting the scoop on exciting new industry products, WC has something for you.
>
> (http://www.wetcanvas.com, retrieved 15 October 2009)

Also see the Artistsnetwork.com at http://www.artistsnetwork.com/ artistsmagazine (retrieved 15 October 2009).

Nevertheless exceptions exist. Cloudscape painting has a website, which presents a brief history of this art stretching back into the nineteenth century (http://www.absoluteastronomy.com/topics/ Cloudscape_%28art%29, retrieved 15 October 2009). A similar site is also available for marine art. It, too, offers a brief history (http://www. absoluteastronomy.com/topics/marine_art, retrieved 15 October 2009), stating that marine art was prominent in England between the seventeenth and nineteenth centuries. It often featured ships and the sea.

Gathering and studying mushrooms

Although numerous edible plants have always been available in the wilds of our planet, nearly all have also been harvested primarily as part of the subsistence diet of people living nearby. Such gathering, even if sometimes pleasant, is nonetheless obligatory. Therefore it would be stretching the concept to call it leisure. Not so, however, for many of those who gather mushrooms today.

Gary Fine (1998), who has conducted the only ethnographic study of people with a leisure passion for mushrooms, classifies them into three types: amateur mycologists (study mushrooms scientifically), hobbyists (compile lists or build collections of mushrooms) and 'pot hunters' (gather mushrooms for eating, for the pot, as it were). The terms 'mushroomer' and 'mushrooming' refer only to the hobbyists and pot hunters and to their activities. Fine also notes that some people are interested in mushrooms as photographic subjects. These amateurs belong in our earlier discussion of plant life photographers.

Fine (1998, pp. 17–19) provides a short history of mushrooming, which begins with the observation that consumption of this fungus undoubtedly dates to the earliest humans. More recently, in 300 B.C., Theophrastus wrote that mushrooms were prized both for trade and as a food. In the latter capacity they were considered delicacies by some. And there is evidence that in Mesoamerica, Siberia, and Scandinavia mushrooms were believed to have both nutritional and psychedelic properties. It was not until the eighteenth century that man began taming the mushroom, which in France was done in caves. By the end of the nineteenth century mushrooming (for the wild variety) had become a leisure pursuit, and a group of French naturalists had formed in 1884 the first mycological society: the Société Mycologique de France. Mushrooming is a well-established free-time activity in North America, though the French and the Eastern Europeans took to it much earlier.

The main type of formal organization of leisure centred on mushrooms is the mycological society. Many countries and continents have a national or international society that welcomes both amateurs and mushroomers. These groups are found, among other places, in Australia, Britain, Africa, Latin America and the United States (including Canada). Many professionals also belong to them. Membership data for these organizations are hard to come by. One of the few to publish membership figures, the British Mycological Society claims 1400 members, many of them international (http://www.britmycolsoc.org.uk/about. asp, retrieved 26 October 2009). It lists 35 local fungus groups located throughout the United Kingdom. The American society does not do this, though local groups organized along state and provincial lines do exist in North America. It is likewise in France, Germany and Italy, among other countries. The larger organizations offer at least one journal to their members as a source of new data and an outlet for publishing research by them and non-members. Some hold conferences. All provide a range of information of interest to amateurs and hobbyists in the field.

Hobbyists go into forests and fields, often in the company of others, to gather or observe mushrooms in what are known as 'forays'. Fine's (1998, p. 97) research suggests that local, regional and national organizations orchestrate many of these events. It is as much a social occasion as anything, but participants also gain considerable information about and experience with identifying mushrooms. And the mushrooms themselves, growing in their native habitat, have enormous aesthetic appeal, though as Fine observes, going into nature to see or collect them can have its drawbacks. There may be annoying insects, unpleasant weather, prickly and sometimes poisonous bushes and so on. Perseverance is a clear requisite in this hobby. Indeed, developing stamina, avoiding natural threats, recognizing species and avoiding those that are poisonous number among the principal challenges of this nature activity.

As in the other sciences where amateurs parallel professionals, the former in mycology aid the latter by collecting descriptive data and making field observations (Fine, 1998, p. 223). Professionals studying an uncommon species may call on amateurs who live near their habitat for detailed observations on them. This arrangement is nevertheless unusual. Instead most of today's amateur mycologists are content to bring to professionals what they first believe are rare or new species. The role of the latter is to identify the specimen, possibly seeing for the first time a new type. Amateurs can also collect professionally valuable information about seasonal occurrences of certain mushrooms, their edibility, and contribute a variety of tips on cooking the fungus. Moreover, from the eighteenth through the early twentieth centuries, it was often the amateurs who named particular species (Watling, 1998).

The wonder for the mushroomer that comes with this NCA is evident in the following quotations:

> Bill Jensen observed that 'Julie [his wife] and I like to hike. One of the joys of mushrooming is being in the wilds and enjoying nature and then getting something [mushrooms] to have with dinner that knocks your socks off.' The 'high' for [Ardean] Watts is identifying mushrooms first for the table and secondly 'because they are beautiful, exquisite organisms,' he said. 'The third thing that hooked me [the author] was the fascinating ecology of mushrooms and their relationship with trees. It is an amazing symbiotic relationship'.
>
> (Prettyman, 2006)

Collecting and gathering natural objects

Let us start by distinguishing dealing, collecting, accumulating and gathering. The range of collectibles is enormous, showing a diversity as wide as stamps, paintings, rare books, violins, minerals and butterflies. With experience collectors become more knowledgeable about the social, commercial and physical circumstances in which they acquire their cherished items.[2] They also develop a sophisticated appreciation of these items, consisting mainly of a broad understanding of their historical and contemporary use, production, significance and, for some, their artistic merit.

Compared with commercial dealers, hobbyist collectors are a different breed. Since dealers acquire their stock to make a living from its subsequent sales, their motives clearly diverge from those driving hobbyist collectors. Although the latter may try to make enough money selling a violin or painting to buy one of greater value, they are usually more interested in gaining a prestigious item for social and personal reasons, or possibly for hedging inflation, than in contributing directly to their livelihood. Additionally, unlike the typical dealer, many collectors hope to acquire the entire set or series of a collectible (e.g., all the posters of the Newport Jazz Festival, all the books in the Nancy Drew series).

The casual collecting of such things as pebbles, matchbooks, beer bottles and travel pennants is, at best, a marginal form of hobbyism. With such items there is nowhere near the equivalent complex of social, commercial and physical circumstances to learn about; no substantial aesthetic or technical appreciation to be cultivated; no comparable level of understanding of production and use to be developed. Casual collecting is therefore most accurately classified as casual leisure, as simple diversion. Olmsted (1991) qualified as 'accumulators' those who collect with little seriousness.

For the purposes of this book gathering refers to acquiring a resource for subsequent use in a hobby, often classifiable as the making and tinkering type. Thus, some people gather driftwood for sculpturing or home decoration. Others gather beach pebbles, sea shells or beach glass (defined below), which, for example, they assemble as mosaics or bottled decorations or for display as individual pieces. Gatherers are hobbyists of the maker and tinkering variety, not collectors (or casual leisure accumulators).

Nevertheless, as shown in this chapter, some activities, gathering, collecting and making a craft form a seamless hybrid hobby. Collecting/gathering/crafting, unlike collecting stamps or fine paintings, for example, is based on a substantial future project of creating an artistic expression or display with the objects collected or gathered. This section concentrates on this hybrid hobby.

Another class of gatherer is excluded, however, mainly because the core activities of their hobby – sculpting, craft constructions – take place in some sort of atelier outside nature, as defined in this book. This is not to deny that nature poses a challenge in an awe-inspiring setting for gathering the raw materials for these activities. Yet these activities are both subsidiary and preliminary, akin to gathering wild blueberries or raspberries for pies; buds, needles or essence of spruce trees for spruce beer; and birch bark for canoes.

Among the chief natural challenges in nature collecting are learning what good specimens look like, where to find them, how to reach them efficiently and how to acquire and preserve them. For example, an experienced collector of, say, grasshoppers, fossilized crustaceans, or conch shells will know to look on certain kinds of bushes, in certain types of rock, or along certain stretches of ocean beach, respectively. Hobbyist collectors also make an effort to learn what science has to say about their collectibles.

Collecting rocks, minerals and fossils

Collecting rocks and minerals goes at least as far back as the days when prospectors were fired by visions of finding precious stones and minerals endowed with great commercial value. Gold and diamonds were, among the various gem stones, the most prized. But, theoretically speaking, the prospecting of yore was not collecting at all, for the prospectors wanted to sell what they found. Nonetheless they helped establish the idea that hunting for particular kinds of rocks and minerals could be fun in and of itself, and so the hobby of collecting specimens of these objects for display gradually caught on.

It should be understood that we are discussing hobbyist collectors, as defined in Chapter 1. The literature on rock collecting, hounding, lapidary work and the like sometimes refer to them as 'amateurs'. Many a rock and lapidary club has amateur in its name. But, as near as we can tell, there are very few truly amateur geologists or mineralogists, as defined in the serious leisure perspective.

Proctor (2004) writes that popular American rock hounding began in the 1930s with electrification and automobility, which enabled enthusiasts to reach the mountains and deserts of the West. Here agates, jasper, petrified wood and other precious stones awaited the collectors and gatherers. Soon rock and gem clubs sprang up around the country, and not long after, manuals began to appear along with specialized equipment for pursuing the hobby. American interest in rock hounding peaked in the early 1960s, when an estimated three million collectors and gatherers were active. The rise of television, exhaustion of easily accessible sites and privatization of land contributed to the decline.

Rocks are often classified according to three types: sedimentary (formed from sediments), igneous (formed from volcanic activity) and metamorphic (sedimentary rock changed by heat and pressure). Each type has several subtypes, and knowing the characteristics of each one helps identify which type it is. The rock hound's (rock collector) goal is to find a unique sample of each rock, which then becomes part of that person's collection. These collectors also read about the geography, petrology and mineralogy bearing on the objects they collect. Note here that, since most fossils are contained within sedimentary rock, fossil hunters amount to a special class of rock collector.

As for the awe of nature felt while collecting, it comes from being in the ambient environment of water, mountains, forest or field. Here is one man's passion and awe:

I have amassed quite a collection of mineral specimens, fossils, gem-stones, crystals and such over the years. I know, some folks go fishing, some go hunting, some are into sports and such, and get a lot of joy and satisfaction from it. I guess it's like the adage 'to each his own'. Well, for me, going on rock collecting trips has been more...I find that the more I learn about rocks, how crystals form, how wood turns into stone, how gemstones get their color, how rocks and minerals form within the earth and on the earth's surface, and the other geo-logical wonders of God's creation, the more I am fascinated and awed. And the more I learn about rocks, the more fascinating they are....

But, to me, rock collecting is different. To me, there's a fascina-tion that comes from being the very first human being that sees the inside of a geode, or finding a fossil of an animal that was alive three-hundred million years ago, or finding a fossilized animal that has been extinct for millions of years, or maybe finding a water-clear crystal of quartz, right out of the red clay that looks like it has been ground and polished by a highly skilled artisan. Or the fascination that comes from collecting many different fossilized sea animals in road-cuts in middle Alabama and Tennessee, showing that at one time that section of what is now the United States was covered by ocean. That's hard to comprehend, but it's true.

(http://blog.greatsouth.net/?p=33,
retrieved 24 November 2009)

Collecting rocks, minerals or fossils are worldwide interests. Many countries and regions publish maps or lists of sites where this hobby may be pursued. And clubs are plentiful in most of them, commonly characterized by a sizeable youth membership. Also common is an interest in the hybrid side of the hobby, in polishing and arranging rocks and stones in displays, adornments, mosaics and the like.

Beach collecting

Serious leisure beach collecting encompasses, in the main, the hobbyist gathering of fossils (i.e., fossilized parts of marine vertebrates), sea shells, sand, and sea and beach glass. 'Sea glass', or 'mermaid tears', refers to shards of coloured or clear glass, often frosted, washed up on the beaches of oceans and large lakes. Often these shards have been repeatedly pummelled and polished by the sand and waves. They originate as litter, primarily through human carelessness aboard boats and on land.

   Sea shells. Consider the hobby of collecting sea shells, as presented by www.seashells.com (retrieved 26 December 2008):

> Walking along the beach picking up seashells and sea life has been enjoyed by millions of people throughout the world. There is nothing more heart-warming than watching a young child run along the beach for hours on end excitedly pointing out all of the fabulous creatures and sea shells that can be found along the seashore, and gathering up seashells to listen into. With so many miles of shorelines throughout the United States and the world, all covered with nature's treasures, it would be difficult not to find this hobby enjoyable. Remember that our coastlines are an important part of the environment, so please try to leave an area as pristine as it was when you arrived, if not more so. Please do not litter and if you see someone else's trash gather and dispose of it properly. If you flip over a rock to look for specimens make sure to replace it when you are done looking. All of these tips will help to keep our shorelines a viable habitat for all.

This site goes on to describe ways to identify and classify the shells found on ocean shores. It groups them under the following headings: molluscs, sand dollars, star fish, sea urchins, sponges, crabs, sea fans and 'other oddities' such as egg casings, seahorses, trigger fish and mermaid purses. This exemplar of non-consumptive leisure grows marginally more expensive with acquisition of one of the many guidebooks for identifying shells and possibly a camera (it may be disposable) for recording the collecting experience and the collectibles found.

Nonetheless, a case may be made for qualifying this serious leisure activity as, in general, essentially non-consumptive. Essentially non-consumptive so long as it remains as collecting, without evolving into, for instance, a making and tinkering hobby (e.g., polishing rocks) or amateur mineralogy (e.g., conducting descriptive analyses on the collected rocks according to established scientific theory and taxonomy). The latter two require specialized equipment, which draw the participant into the marketplace and the expansive sphere of consumptive leisure.

Sand. The International Sand Collectors Society (ISCS) – founded in 1969 – maintains that people have been collecting sand for over a century (http://www.sandcollectors.org, retrieved 30 November 2009). It is indeed an international hobby, as seen in the numerous links listed in the website 'Sand Collecting' (http://www.fernwood-nursery.co.uk, retrieved 30 November 2009). It offers a quarterly newsletter – *The Sand Paper* – to members. Another website (http://www.sand-collecting.org), in its overview of the hobby of sand collecting, describes the nature, joys and location of this pursuit as well as the motives for engaging in it:

> Sand collecting is based around the simple idea of visiting a variety of different locations such as beaches, rivers, lochs, quarries and so on, and collecting one or more small difference samples of sand from each. The samples are then prepared and displayed at home, usually inside small clear plastic or glass containers or bottles.

> People collect send for many reasons. For some it's simply a great excuse to get outdoors and visit new places, while for others it's just the many different colours and textures of the sand that attracts them but for many, myself included, it's really the joy of finding new and interesting samples of sand – and adding them to the collecting.

> Collecting sand is one of the easiest and least expensive hobbies you can get, all you need are a few self-sealed polythene bags to collect the sand in, an old spoon to scoop the sand up with and some clear containers to house your growing collection....

> Sand collectors, sometimes also known as Arenophiles or Psammofiles, differ in what they actually collect, some collect only local sands, perhaps from their own county or country. Others collect sands from all over the world, either by traveling to different countries themselves

or by exchanging sand samples with collectors from other countries. There really is no limit to how many samples you can collect.

Sand collecting is not limited to beaches, as its treatment in this section might suggest. Still the beaches of oceans, lakes and rivers may well provide the most awe-inspiring environments in which to collect. Moreover some collectors, to complete their collections, may buy otherwise unobtainable samples, offered for sale over the Internet and through newsletters.

Sea and beach glass. According to the Glass Beach Jewelry and Museum, sea glass is found everywhere, even while some areas are especially renowned for it (e.g., Hawaii, Southern Spain, Northeastern United States). Glass making began in the Roman period, but just when some people began hobbyist collecting of glass shards washed up on beaches remains unknown. Glass Beach's website states that the value of these artefacts

> is partially determined by [their] color. This is because only a few items were stored in red, blue, lavender, purple or pink glass containers. Likewise certain rare tints and shades of these popular colors are found.... For instance, very rare Cobalt Blue, the 'sapphire' of the beach, came from such apothecary items as Milk of Magnesia, Vick's Vapo Rub, Noxema, Nivea, and Bromo Seltzer bottles, along with some prescription bottles and perfumes.
>
> (http://glassbeachjewelry.com/history.htm, retrieved 27 November 2009)

Some serious leisure collectors, having learned what to look for, use the shards they find to make crafts such as jewellery and mosaics. Others fill vases or bowls in their homes with the pieces, using them as decorations. At the casual leisure level people collect glass as part of a session of beachcombing, which they do for relaxation. Serious collectors gather sea glass whose colours appeal to them, often spurred on to find their favourite shades. Many want to expand their collections. Some are in search of a piece of sea glass in a rare colour. Since bottles are the most common source of sea glass, white (clear), brown and green – the colour of bottles produced during the past 40 years – are prevalent colours. Orange is the rarest colour, and red, turquoise, yellow, black, teal and gray are extremely rare.

The North America Sea Glass Association (established 2006) holds an annual festival that draws 4000 people and circulates a newsletter to 1300 members. It also sponsors a blogging service. The Mermaid Tears Sea Glass Festival of Prince Edward Island, Canada, was held for the first time in 2009. Festival chair, Nancy Perkins, began collecting sea glass in 1981. Perkins's initial find was a rare red piece, after which she was hooked. 'The fun of going to collect it is wonderful', Perkins said. 'We walk along the shore, and it's usually in a quiet place, so we hear the birds and we see the seals.... We find little hidden spots that we wouldn't have found if we hadn't been looking for the sea glass' (http://www.cbc.ca/canada/prince-edward-island/story/2009/ 07/17/pei-sea-glass-festival.html, retrieved 27 November 2009).

The website Odyssey Sea Glass (http://www.odysseyseaglass.com/ index.html, retrieved 27 November 2009), based in Washington State, offers information on good collecting beaches in North America, Britain and the Caribbean. It also has an online newsletter aimed at an international readership. It carries information on sea glass collecting, sales (of rare specimens to collectors), literature and festivals available anywhere in the world. The Sea Glass Network in Australia performs similar functions, often linking to relevant events elsewhere in the world. The social world of this hobby is further enriched with several recently published books and locally organized workshops. Its periodicals appear to be entirely of the newsletter variety. Finally, as with rock collecting, there is also a hybrid side to this hobby, pursued through arranging colourful shards in displays, adornments, mosaics and similar crafts.

Amateur botany

Serious leisure participants in this science contribute to it primarily at the descriptive level, as by mapping local distributions of certain plants, trees, shrubs, flowers and grasses and by identifying new species. So, in June 2008, David Gowen discovered two new related wildflowers in the phlox family growing in the Walnut Creek area of California (Crooks, 2008). He is among a handful of amateurs to contribute this way in recent years to his science.

The science of botany goes back a long way. Classification was attempted in ancient India and China, and in the latter, a body of knowledge accumulated around the medical use of plants. But botany as a pure science began in the fourth century B.C., when the Greek philosopher Theophrastus wrote on classification. The Oxford Botanic Garden was founded in 1621. Few people if any made a living from botany in those days; these scientists were amateurs, or more accurately, hobbyists (Stebbins, 1992, pp. 8–10). Keeney (1992) holds that, by the end of the nineteenth century in America, it was common for this field to distinguish amateurs from professionals.

Jason Hollinger's website (http://alumnus.caltech.edu/~hollin/index. html, retrieved 30 November 2009) contains a list of the equipment that amateur botanists commonly need. Purchase of a dissecting scope and a sophisticated technical flora (much more detailed than a Peterson's guide) raise significantly the cost of pursuing this hobby. To these expenses, we must add those of suitable clothing and transportation. Most botanical societies in Europe, North America, and Australia and New Zealand welcome both amateurs and professionals to their ranks. Members typically receive a scientific journal, a newsletter and information about related publications, websites, conferences and so on.

As for numbers of amateur botanists precise figures are difficult to obtain. According to their websites the Botanical Society of South Africa, for example, has 15,000 members and the German Botanical Society around 900. These organizations, like their counterparts in other countries, welcome both amateurs and professionals, with significant numbers of them living outside the country of incorporation.

Fauna

The zoological class of fauna is the class of animals. We use the broad definition of animal here: a living organism having sensation and voluntary motion, without rigid cell walls, and dependent on organic substances for food. Thus, a fish is, by this conceptualization, an animal, a faunal creature.

Fishing

Fishing as an NCA consists, in the main, of finding and catching with a rod and reel, spear, jigger or bare hands (e.g., catfish 'noodling') the fish inhabiting rivers, creeks, lakes and oceans. This is sport fishing. Netting fish, trapping lobsters, digging clams and so forth are, as leisure, gathering activities; participants seek resources for making and tinkering hobbies like cooking and preserving. For reasons given in the preceding section, such gathering is not covered in this book.

The challenge nature sets in the sport, or recreational, fishing hobby is navigating or walking in the water being fished (e.g., in waves, depths, rocks, currents, weed beds), presenting some bait (including artificial lures) such that fish will bite, and reeling in and possibly netting those that do. Successful fishers know how to navigate or walk the waters they fish (or walk on and drill into ice for ice fishing). They are skilled at presenting bait, setting hooks and reeling in their catch, the accomplished fly fisher being the leading exemplar here. They are also knowledgeable about the baits used in different meteorological conditions. Some practice catch-and-release, others 'meat fish': motivated in good part by a need to supplement the larder of family and friends.

Whereas man has always consumed fish as a food, sport fishing is of far more recent origin, though as Yoder (2004a, 2004b) observes, it is impossible to determine when. Nevertheless he notes that Englishman Izaak Walton wrote *The Compleat Angler* in 1653, and that the British passion for freshwater fishing eventually came to America, becoming popular there by mid-nineteenth century. Saltwater, or big game, fishing has a similar history in Australia, New Zealand and the United States, it being part of their British heritage. In the twentieth century an important part of the history of this hobby has revolved around improvements in equipment, notably rods, reels and electronic 'fish finders'.

Many sport fishers are unorganized. Having no desire for club affiliation, they fish alone or with friends and relatives. Some prefer that these be same sex occasions. The unaffiliated fishers may well subscribe to a magazine (e.g., *Bassmaster, Fly Rod and Reel, Marlin*) or browse a website (e.g., Sea Fishing, The OutdoorLodge.com, BigFishTackle.com forums) centred on their kind of fishing and patronize from time to time the local and online stores that sell the tackle, bait, clothing and accessories they need. There are also television programmes, among them, 'Adventures in Fishing' (USA), 'Saltwater Ventures' (USA, Caribbean, Central and South America) and 'Fishing Australia'. Some countries and regions hold annual fishing shows and exhibitions, where tackle and accessories are displayed and tournaments and other competitions may be held, a tradition well exemplified in bass fishing (Yoder, 1997).

Others enlarge this social world by joining a formal organization. Salt water and freshwater fishing clubs are found throughout the world, many of them oriented by member preference (Bryan, 1979) for capturing certain species, sometimes limited to certain kinds of tackle or lures (e.g., fishing trout with dry flies only). As for freshwater fishing, the website of the Ultimate Bass Fishing Resource Guide lists hundreds of clubs in North America as well as some in Africa, Spain, Portugal, Japan and Italy. Trout fishing clubs exist in Britain, North America and New Zealand, among other places. An advantage of belonging to one is receiving information about such matters as fishing tackle, techniques, productive fishing sites, fishing competitions and clothing and equipment. Books about fishing – and there are many – are commonly advertised and reviewed in the periodicals of the larger clubs. Above all the clubs offer at their meetings an opportunity for exchanging fishing stories, receiving advice and possibly meeting someone to fish with.

Data from the United States suggest that sport fishing is a highly popular hobby. In 1995 there were 10.5 million recreational saltwater fishers and in 1997 around 39 million participated in recreational freshwater fishing (Kelly & Warnick, 1999, pp. 115–118).

Hunting

For our purposes we classify the quarry hunted in an NCA as: fowl, large animal and small animal. Thus, people in diverse parts of the world hunt, for instance, ducks, geese, pheasants and turkeys, as well as deer, moose, elk, bear, lions and elephants. Among the small animals (and rodents) hunted are foxes, rabbits, squirrels and prairie dogs. Hunting any of these creatures as pests is not a leisure activity, but an obligation (work or non-work), even while it usually presents a challenge and might at times be agreeable.

According to Hummel (2004) recreational hunting appeared in the United States in the early nineteenth century, the shooting before then of fowl and animals being done largely for subsistence. Nonetheless hunting for fun among the general population arrived earlier in America than in Europe, where until late in the nineteenth century, only the elite were legally permitted to engage in it. By mid-nineteenth century middle-class American hunters were forming clubs and leasing land as private hunting grounds. These organizations have now for the most part disappeared. Today the hunting tradition persists in rural areas and small towns located near wildlife habitats, though recent evidence suggests renewed interest among some city dwellers (Farrell, 2009)

The natural challenge to be met in hunting is above all marksmanship, skilfully using a gun or bow to kill the target. Other challenges include tracking it or successfully disguising oneself (e.g., using duck blinds, tree stands, camouflaged clothing, deceptive sounds) so as either to fool or to lure game into coming within range of the weapon. Moreover considerable knowledge is required to carry off a successful hunt, one major component of which is an understanding of the habits of the quarry in diverse climatic conditions. Another is familiarity with equipment, guns, bows, camouflage, trucks, boats and motors (for some duck hunters) and so on. An ability to read maps and global positioning system (GPS) devices may be an asset. How to clean and dress the killed bird or animal is also part of this know-how.

As with fishing some people hunt alone or with friends and relatives – often preferring the same sex – and thereby remain outside formally organized hunting. In this regard some types of hunting require team effort, as in hunting antelope and wild ass in Iran (Mahdavi, 2007) and moose in Norway (Sande, 2001). Unaffiliated hunters may also subscribe to hunter's magazines (e.g., *Wildfowl, African Hunter Magazine, Australian Hunter*), browse relevant websites (e.g., Hunter-Information.info, Supportfoxhunting.co.uk, Global Sporting Safaris), frequent local and online hunters' stores, watch televised programmes (e.g., WildTV, The American Outdoorsman, Los Gauchos Television), attend shows and exhibitions devoted to hunting and the like. Moreover, as with fishing, the social world of hunters can be expanded by joining a club. Here they gain the same kinds of benefits the fishers do. Kelly and Warnick (1999, p. 107) reported that, in 1997, 17 million Americans were hunters.

The joy of being in nature while hunting would be, you might think, self-evident. Still Hunt et al. (2005) found that only part (no proportion stated) of their sample of moose hunters in northwestern Ontario in Canada also hunted for this reason (see also Reid, 1991). Many members of their sample are interested primarily in finding a quick and easy way to bag their moose, the natural environment in which this takes place having little or no special appeal.

The activity of trapping wild animals for their fur or food has intentionally been excluded from this section on hunting. Trapping is a skilled undertaking, but it appears, one mostly done not as leisure but as a livelihood: the trapped animal's pelt is sold or its flesh is eaten, if not both. In the past some native peoples also used the pelts for clothing or shelter. In the language of the serious leisure perspective, modern trapping is either work – possible only during legally-established seasons – or non-work obligation – often to supplement the family food supply.

Bird-watching/amateur ornithology

Bird-watching, or birding, consists of the hobbyist activity of observing and listening to birds in their natural habitat, typically using the naked eye, binoculars or a spotting telescope. Many birders also carry a field guide and a notebook. For most keep lists of birds seen and heard and where this happened. Nonetheless not all bird habitats exist in nature as we have defined it, as attested by the great range of city birds and birding equipment (e.g., birdbaths, birdhouses, squirrel-proof bird feeders). So only some bird-watching can be labelled an NCA.

David Scott (2004) writes that birding and collecting were, until the late 1800s, synonymous. Before this ornithologists and bird-watchers alike would shoot birds and collect their eggs and nests for scientific study. A combination of improved optics and a concern for declining bird populations put an end to this practice. The invention of binoculars with prisms made field identification possible without killing the subject. Scott (2004) reports survey estimates from 2002 that 46 million Americans are bird watchers, though only 40 per cent observe away from home.

An enthusiast need not join a club to watch birds or pursue amateur ornithology. Indeed too many people on the ground could spook the birds being observed. Nonetheless bird clubs and associations abound the world over, lists of which are available on a variety of websites where they are classified by nation, region, species and other criteria. The Audubon Society, with its local, regional, national and international wings, is arguably the most prominent of them. Through their own websites and newsletters, these organizations typically provide birders with information about birding sites (many are international), bird counts, tours, advice, guides, observing equipment and so on. There are also competitions to determine who can count the most species in a specified period of time. Amateurs often join these clubs as do bird owners of, say, parrots and canaries. Owners, as such, are not birders, and the multitude of bird shows held around the world, often sponsored by bird clubs, are mainly staged for people interested in buying a pet. Bird clubs therefore also provide information on caring for these pets and avian pet shops comprise part of the owners' social world. Meanwhile the NCA bird-watcher has access to numerous magazines, including *Birdwatch, Wildbird* and *Dutch Birding*, and various websites, among them, Birding-Aus (Australia), Euroblrdlng.com and Clare Birdwatching.com.

Diane Porter describes her delightful walk in a forest during spring where, by chance, she encountered two woodpeckers.

Storm of Woodpeckers

In the lovely, warm, late April afternoon, Michael and I walked into a woods near where we live. Little tender leaves were showing against brown branches. Violets were under foot, along with the fuzzy white heads of pussytoes, my favorite early spring flower.

We picked our way, clambering uphill through tangled second-growth woods. And then at the top of a knoll, suddenly the walking was easy, the floor of the forest open. We had entered a stand of ancient oaks, somehow spared when the original forest fell to the saw a hundred years ago....

Close to the path was a tall snag, a tree that had been killed by disease or lightning. All its branches were gone, and most of the bark had sloughed from the still-standing trunk. While Michael and I stood there talking about the nutrient cycles of the forest, a male Red-bellied Woodpecker landed on the snag near the top and called a strident 'Chirrah! Chirrah!'

Immediately a female woodpecker poked her head out of a hole half way up the dead tree, cocked her head to look up at him, and came out and clung to the trunk.

The male lit by her side for a moment and then entered the hole, while the female flew into the woods. We knew what this meant. They had eggs!

Michael and I sat down and watched. The birds took turns inside, never leaving the nest unattended for more than a minute. As for our intrusion, after taking one long look at us, the birds seemed to ignore us. We were delighted.

> (http://www.birdwatching.com/stories/red-bellied_
> woodpecker_nest.html, retrieved 5 December 2009)

Amateur ornithology

Amateur ornithology is a branch of the scientific study of birds. Birders thrill at watching birds, especially species rarely seen, observing their actions, and learning what science has learned about them. Amateurs may have similar interests, but they also have scientific ones. Ainley (1980), in her study of the rise of professionalism in American ornithology, says that initially this science was considered part of the study of natural history. Scientific work on birds by gentlemen-naturalists in the United States dates to the early nineteenth century and consisted mainly of collecting and classifying them. By mid-nineteenth century ornithology had become a part-time occupation for a few scientists, some of them having been assigned by government to accompany exploratory expeditions into the western regions. But it was in the last two decades of that century, with its increasing number of paid teaching and research positions in the science that, the distinction between amateur and professional became necessary.

Amateurs participate extensively in leg banding (to trace migration habits, territoriality), bird art and photography (to assist recognition), periodic bird counts (to learn about carrying capacity of habitats) and recording bird songs. In short, during this leisure, they also experience nature. In addition they often write in the professional journals. Ainley found that 12 per cent of the articles published in 1975 in three leading ornithological journals were penned by amateurs. The organizational world of the amateur in ornithology is much the same as that of the hobbyist birder, the principal exception being the ties of the first to the social world of the professionals, to their conferences, journals, websites, research projects and, in general, their intellectual milieu.

Collecting insects

Vladimir Nabokov, a Russian writer and part-time entomologist, once observed that he could not 'separate the aesthetic pleasure of seeing a butterfly and the scientific pleasure of knowing what it is' (reported in Boyle, 1959). In fact man has had either a fascination with or a fear of insects stretching far back into history. From time immemorial people have marvelled at the actions and activities of local insects, while learning to fear or loath some of them. Thus certain species became objects of literature, theatre, painting, and later, music, cinema and photography. In some cultures myths grew up around particular insects, not infrequently spiders. In the United States hobbyist collecting and the amateur study of insects was pursued from the mid-nineteenth century to the first two decades of the twentieth century in parallel with the rise of professional entomology (Rogers, 1960).

David L. Keith and Tiffany Heng-Moss, both professors of entomology, offer detailed advice on collecting and preserving insects (http://entomology.unl.edu/tmh/cnt115/labs/collecting.htm, retrieved 24 November 2009). For example collectors should examine plants closely, looking for holes in leaves or ragged leaves near the ends of branches. At minimum collectors in the field need: 'an insect net, one or more storage boxes, insect nets, pins, pinning blocks, spreading boards, light and pitfall traps, killing jars, killing and preserving chemicals, several vials of assorted sizes, plastic bags and assorted containers'. Insects must be killed before pinning and mounting. The killing jar contains a toxic atmosphere created by a liquid fumigant or killing agent. The goal is to kill collected insects as quickly as possible.

Preserving techniques vary according to whether the insect has a hard or soft body. The first type requires little maintenance, whereas the second must be preserved in liquids inside rubber-stoppered glass vials. Hard-bodied insects may be pinned for display, using special techniques for pinning different species so that body and wings are as fully visible as possible. Collections are commonly arrayed in 'Schmidt boxes', where they are labelled by entomological order, location and date of finding and name of collector. Neatness in pinning, mounting and displaying of specimens is prized.

Rare insects and those from distant parts of the world may be purchased from dealers or acquired by trading. Some noted insect collections have been sold at auction. Insect collecting may take place either in the wilds, qualifying thereby as an NCA, or in man-made areas such as private gardens and public parks.

Local entomology clubs operate in most parts of the world. Some have been established for and are run by university students, the majority of whom are seeking an education in entomology. The 4H in a number of American states sponsors such clubs for youth. Local clubs also hold exhibitions, or fairs, one main attraction being displays of personal collections for viewing by other collectors and the interested public. Some clubs attract photographers interested in insects as subjects. National and regional entomological societies are active on every continent, with many welcoming both amateurs and professionals. Some encompass all entomological interests; others specialize by species of insect.

Conclusion

The floral and faunal hobbies and amateur pursuits are, in general, quite environmentally friendly. Painting and photographing nature do literally no damage to it. The same may be said for collecting objects, including taking living ones such as leaves and mushrooms, though this assumes that the organism thus modified can regenerate itself.

Faunal pursuits in general have a somewhat less enviable record of environmental responsibility than the floral set. True fishers and hunters who respect the legal limits of their catches, assuming those limits are designed to sustain the species in question, help ensure that the species will remain healthy and abundant. In another approach to sustainability of a species, these limits may be raised or lowered, sometimes annually, to help prevent its overpopulation or predation on domestic animals (e.g., Herfindal et al., 2005; MacMillan & Leitch, 2008). Poachers, depending on how active they are, would upset unfavourably this delicate balance of leisure and nature, as does some trophy hunting

(Coltman et al., 2003). As for bird watching and amateur ornithology, both are like nature painting and photography in that the former two activities leave no permanent footprint on the environment.
Nevertheless Theodori et al. (1998) found that pro-environmental attitudes and behaviour correlate positively with education, liberal political
views and participation in outdoor recreational activities and not with
whether the activity pursued is 'appreciative' (e.g., bird watching, nature
photography) or 'consumptive' (e.g., fishing, hunting).

And it is virtually the same for the collection and scientific study of
insects; some insects are taken, to be sure, but not enough to upset their
role in the balance of nature. Furthermore, there is little danger that collectors and amateur entomologists will harm nature, unless they destroy
too many insect homes by, for instance, forgetting to replace the logs
and stones they have turned over in their searches.

Of the activities discussed in this chapter, only the collecting of
natural objects and the gathering of mushrooms can be qualified as
essentially non-consumptive leisure. But, even here, the costs of transportation and clothing cannot be ignored. In the other activities the
outlay for equipment looms large, as exemplified in guns, cameras, oil
paints, microscopes and fishing tackle. Furthermore the overall expenditure for any serious leisure activity can be increased significantly by
extensive participation in its social world, as through membership fees
in societies and websites, magazine subscriptions, costs of books and
manuals, and transportation and other expenses related to monthly club
meetings and annual conferences.

Note, however, that some of these purchases enabling pursuit of the
hobby or amateur activity are one-off acquisitions. It may cost plenty
to buy a decent microscope, shotgun or SLR camera, but it will last and
an economy-oriented participant will be satisfied with it. From here on
that person's NCA is virtually non-consumptive. This observation, of
course, assumes such people can resist temptations to replace equipment
that notwithstanding commercial hype fail to augment noticeably their
capacity to engage in their leisure passion.

As with many other NCAs clothing can be a main consumer outlay in the floral and faunal activities. First, boots, gloves, jackets, hats
and the like do wear out and must be replaced. Second, new materials and designs are seemingly always being invented and marketed,
with some being truly advanced over their predecessors. They can genuinely add to the comfort and effectiveness of the pursuit. Thus, since
the 1970s, Gor-Tex fabric has become renowned in outdoor clothing for

its combination of breathability and water repellency. It has undergone major improvements over the years, but for all that, is now facing challenges from newer fabrics such as eVent and Epic. An NCA enthusiast might well want to spend considerable money to buy clothing enabling him or her to pursue the activity with significantly greater ease and success.

6
Ice and Snow

For some nature challenge enthusiasts ice or snow (sometimes both) have significance far deeper and more positive than the common view of both as something that precipitates a fall or requires cleaning up. What is more the latter, negative, appreciation of ice and snow is usually accompanied by a dislike for the freezing temperatures that make both possible. And the disagreeable experience of freezing temperatures may be worsened by wind, producing painful 'wind chill'. So the impulse, among those who can afford it, is to escape to a place on the planet where these conditions are absent and agreeable warmth, free of ice and snow, prevails.

But one man's poison is another's meat. A hearty minority of people find great self-fulfilment in the very elements that the people above reject with equal fervour. Ice or snow, to be carefully watched to be sure for the accidents they may trigger, are nonetheless the foundation of a variety of winter hobbies and sports. *Homo Otiosus* has been admirably inventive when it comes to ways to enjoy himself on these two natural surfaces. To some extent the results have been inventions of necessity. Stuck in a winter climate for several months of the year, it can be dispiriting trying to wait it out indoors. An antidote: look for something to do outside that is enjoyable, if not fulfilling. Moreover, history shows that some of these NCAs originated as aids to work and subsistence.

This book centres on NCAs. Hence, as pointed out in Chapter 1, not all winter outdoor activities fall within its scope (e.g., down-skiing at resorts, ice hockey on artificial rinks). The activities properly classified as NCAs, all but two of which are examined in this chapter, are

- **snow:** skiing (cross-country, back-country), snowshoeing, snowboarding (back country), skijoring, snow kiting, snowmobiling and dog sledding;

- **ice:** ice skating, motorized ice racing, ice boating, ice climbing (covered in Chapter 4 with mountain climbing), ice kiting (covered with snow kiting) and ice fishing (covered in Chapter 5 with fishing).

In each of these we will start with a description of the hobby, its origin and where on earth people pursue it. This section will flow into a discussion of the basis for its classification as an NCA. We will then describe the awe and wonder of nature that comes with practising each in its natural setting. Finally, since some NCAs are also pursued as sports, this facet of the hobby will also be considered. The conclusion, as has been the practice in preceding chapters, looks into the implications of these NCAs for consumption and sustainability.

Snow

We begin with a set of fully self-propelled activities; they share the sensation of enabling full human agency in executing them. Here participants are responsible for their own locomotion, and as a result find fulfilment, in part, in their ability to achieve this. Here they are also free to explore the surrounding environment by going off-trail, listening to birds, watching wildlife and pondering the scenery. These activities are Nordic skiing, snowshoeing and snowboarding.

Nordic skiing

Nordic skiing, which includes the cross-country and backcountry forms, has been observed among the Sami, Norwegian, Swedish and Finnish peoples from c. 2500 B.C. to the present (Crystal, 1994, p. 1020). Originally the two had the same function: an efficient mode of transportation on snow without special machines to prepare a track leading to the skier's destination. The inefficient alternative, for this person, was to trudge on foot through thigh- or waist-high snow. Only as skiing evolved into a hobby did the cross-county/backcountry distinction develop. Some enthusiasts came to prefer what is known today as 'track skiing', propelling themselves on skies, aided by poles, along tracks and packed trails established by either a machine or other skiers. Others favoured off-track experiences in the 'backcountry', which sometimes included in hilly and mountainous areas downhill skiing over unprepared fields of deep snow.

Cross-country skiing on prepared tracks and trails is found in most countries with an annual snowfall of depth and duration sufficient for the activity. Local, regional, national and international competitions are common today, the ultimate being the winter Olympic Games. Still many cross-country and all backcountry skiers pursue the hobby purely for the natural challenge it sets for them. What is that challenge?

The challenge is met in the core activities of this kind of skiing. Cross-country or backcountry, skiers here must develop the skill of pushing themselves on their skies through snow or over a packed track. They accomplish this by 'kicking', or forcefully pressing downward on the middle of the ski, which is cambered, to contact the snow beneath, the middle being coated with wax or having a manufactured rough surface, either of which causes the ski to 'grip' the snow. The skier then terminates the push and glides forward on the slippery tips and tails of the two skis, which have a different wax or surface that facilitates this movement. Using poles further advances forward movement. Some skiers, depending on the gradient, kick and glide on downhill and uphill segments of the course. But many skiers, especially on steep sections, simply glide down the hill or even partially brake themselves with a technique known as 'snowplowing'. Steep uphill sections often require a special walking action called the 'herring bone' step. All this is referred to as 'classic' cross-country skiing, and is now contrasted with the newer form of ski-skating (modelled after ice-skating).

Intermediate and especially advanced cross-country and backcountry skiers have more or less mastered these and other techniques and background knowledge (e.g., about waxing, snow conditions, clothing, food, hydration) needed to find the deepest fulfilment in their core activities. The sensation felt by competent skiers is one of nearly effortless striding and gliding over the snow through a natural winter environment at the pace similar to slow to moderate jogging. That environment, which has an outdoor quietness unique to heavy concentrations of snow, typically consists, in the main, of some combination of snow-laden coniferous and deciduous trees, snow fields, open or partially frozen creeks and rivers, and in some backcountry terrain, steep hills and valleys to climb and descend. Descending slopes covered in deep powder snow generates flow as the skier wrapped in the three-dimensional sensation of descent manoeuvres in and out of a series of telemarks (turns) heads for the bottom of the hill. In both kinds seeing wild birds and animals while en route is considered an additional attraction.

Kate Carter, editor of *Vermont Sports Today*, writes on the fulfilment gained from cross-country skiing.

> The feeling of gliding is what I love most about cross-country skiing. Gliding across a long, flat stretch, keeping my momentum going with

long, relaxed strides and easy double-poling is sensational. Even more thrilling is gliding at high speeds downhill, mastering the forces of gravity while unleashing the flow of adrenaline. Most exhilarating, however, is the sensation of gliding uphill, actually defying gravity and skiing up hills as if they were flat.

I know I am skiing well, that I have gone beyond shuffling, when I am gliding up a hill. It doesn't happen often, but once in a while everything comes together – a fresh dusting of powder, the right wax, proficient technique, intuitive balance – and not even gravity can hold me back.

(http://www.jackrabbits.ca/resources.asp,
retrieved 21 January 2008)

The sporting side of cross-country is concentrated in track skiing, where contestants race over a prepared course for the best time. Flow and exhilaration are maximized here, as the skier fired by competing against others strives for his or her top speed under all conditions of the trail and all aspects of its geographic layout (turns, hills, bumps, etc.), strides and glides or skates over the snowy course. Like competitive running and cycling, races for skiers may cover short or long courses (the Vasaloppet in Sweden covers 90 km), attract serious or casual participants (both sometimes allowed in the same race) and lead to minor (e.g., a plate of cookies) or major prizes (e.g., a solid gold medal).

Snowshoeing

The use of snowshoes may be as old as that of skis, even if the place of origin was different. Osgood and Hurley (1975, p. 11) note that a prototype of snowshoe was developed in Central Asia about 4000 B.C. Migrants from there eastward across the ancient land bridge over the Bering Strait stayed with the snowshoe. Those heading west to Scandinavia adopted the ski. It was thus that the snowshoe became the efficient footwear used by the native peoples of North America for getting around in winter, an artefact they continued to develop. Early versions were made of hard wood frames turned up at the front, to which bark and branches and later rawhide leather thongs were attached in criss-cross fashion (the 'latticework') to provide 'flotation' in the snow. A leather harness forward of centre fixed to a cross bar secured the user's foot. Since snowshoes at this time were conceived as an aid to such work as winter trapping and hunting, the hands of the user had to be free for carrying related equipment. Hence poles were not typically part of the native snowshoer's gear.

Since snowshoeing is fundamentally walking in snow, it is a less complex core activity than cross-country skiing and therefore easier to learn. Nonetheless novices must learn to adapt to frequent unevenness found on both packed trails and off-trail. Skill is needed in ascending and descending hills, whether on packed surfaces or powder snow, and in backing up. Because of the width of the snowshoe, the users' gait is semi-bowlegged, which necessitates some adjustment. As a result of these requirements, snowshoers, even experienced ones, often encounter muscular stiffness after their first couple of outings at the start of the season.

The modern snowshoe is a substantial advance over the old native wood and leather implement. The modern version is made of aluminium, stainless steel or plastic with a neoprene, nylon or polypropylene decking (replaced the latticework) and harnesses as well as a metal footplate. The most recent versions also have crampons along the frame and under the footplate, and are narrower than their wooden predecessors. Moreover many a modern snowshoer walks with poles to aid balance and enhance forward movement. Needless to say the new snowshoes cost much more than the wooden ones, which are still available in certain specialty shops. The former are, however, much easier to maintain than the latter.

Snowshoeing is thus serious leisure, part of which is conditioning. Unless the pace is very slow, an outing lasting more than an hour is enervating for most people. This is especially true if the participant is breaking trail compared with walking on a prepared trail (usually made with a snowmobile) or one over which others have recently gone. However rapid the pace it still amounts to a walk for most hobbyists. As a result there is ample occasion to view nature of the sort described for cross-country skiing. And, as with the skiers, this is a main attraction for the snowshoers.

Three women report their snowshoeing experience in Rocky Mountain National Park in the United States.

We decided to take the Glacier Gorge trail to either Mills Lake or the Loch since we'd been to the Bear Lake area a few times before and most other parts of the park are inaccessible in the winter.

This was an unbelievable trek – both in scenic beauty and physical exertion. We climbed up Alberta Falls, which was totally frozen and buried under feet of snow. We then wound our way along a ridgeline, up another waterfall and frozen creek, and eventually reached Mills Lake, which was a spectacular destination – totally frozen and snow-covered, and ringed with a massive wall of mountains anchored by 14,261' Long's Peak....

It was getting deep into the afternoon so we made record time on the way back, sliding down waterfalls and running down hills through waist-deep powder. After the 6+ miles through deep snow over mountainous terrain, we were absolutely kicked and more than happy to accept Desiree's parent's offer of dinner and lodging for the night. Long story short, snowshoeing is awesome. Let's hope we can get one more in before the melt.

(http://obliviondiaries.com/2006/03/29/
i-love-snowshoeing, retrieved 17 January 2008)

Todd Callaway, manager of a sports equipment store, observes that:

snowshoeing is really easy. If you can walk, you can snowshoe. And it's something that families can do together.

[Still] there is no typical snowshoer. You see people who are into winter mountaineering, winter camping, hiking, running ... all levels, all abilities, all looking for different experiences.

But the biggest converts to snowshoes are families discovering the sport can bring them together.

Families often splinter on the ski slopes. The kids go one way, Mom goes the other way and Dad goes somewhere else. Then, they all meet for lunch, and separate again.

To many people, snowshoeing is freedom. Many people hike all summer and fall, but when winter comes, they can't access the trails they like. With snowshoes, they can go anywhere in the winter that they went in the summer, and it can be an even better experience, because they'll usually have the woods to themselves....

Perhaps the best thing about snowshoeing is the sound: The thrump, thrump of the platter-shaped shoes hitting fresh powder. The rhythmic, frosty breathing from the effort. The errant plop of snow from an overladen tree branch.

Or maybe the best thing about snowshoeing is that it can take you where there are so few sounds - a nearly silent forest in the winter.

(http://findarticles.com/p/articles/,
retrieved 17 January 2008)

Nowadays competitions on snowshoes appeal to a small proportion of all snowshoers. Yet there are many races. *Snowshoe Magazine* listed scores of them for the first five months of 2008 held at various locations inside and outside North America. In the past, especially in Québec, races were popular, some of them even run over track and field hurdles. Osgood and Hurley (1975, pp. 120–125) describe various games played in the past on snowshoes, including baseball, dodge ball and field hockey. How often these games are played today is unknown.

Snowboarding

Snowboarding, which has been described as 'surfing on snow', involves an individual descending a snow-covered slope while standing sideways on a lightweight board about 150 cm long (about 5 feet) attached to the feet. It borrows techniques and tricks from surfing and skateboarding, and can be done wherever alpine and backcountry skiing are possible. Since no poles are used in snowboarding, it is difficult to traverse flat areas. Instead, the ideal terrain is a slope covered in deep, loose snow. However ideal the slope, some snowboarders find flow and fulfilment simply from successfully riding down it, whereas others (on packed runs) also attempt various 'tricks' en route to the bottom. Discussion here is restricted, as indicated earlier, to natural settings for snowboarding, to 'free riding', even if most of the time the hobby is pursued at alpine ski resorts in far more artificial conditions (Stebbins, 2005a, pp. 60–61, describes this aspect of the hobby).

Snowboarding requires a board, some bindings and a pair of boots. Snowboard lengths vary according to size of rider and type of riding that person does. Unlike skiers, who shift their weight from ski to ski, snowboarders shift their weight from heels to toes as well as from one end of the board to the other. When snowboarders shift their weight towards the nose, or front of the board, the board heads downhill. When they shift their weight towards the tail, or back of the board, they head uphill or slow down. Riders execute quick turns by pulling the back foot forward or pushing it backward to change direction. They can stop the board's motion by pressing heels or toes down hard, thus digging the edge of the board into the snow.

The hobby originated in the United States between the late 1960s and early 1970s. It was developed independently by three Americans: Tom Sims, Jake Burton Carpenter and Dimitrije Milovich. Sims is often credited with building the first snowboard, when in 1963, he modified a skateboard to slide on snow, an idea influenced in part by his experience as a surfer. Carpenter tinkered in the late 1960s with a snow toy to which he gave the name 'Snurfer' (it even had a rope attached at the front) and only later realized how ski technology could improve snowboarding.

Milovich, an East Coast surfer, got his inspiration from sliding on snow on cafeteria trays and based his snowboard designs on surfboards.

Snowboarding appealed initially to a small group of surfers, skateboarders and backcountry enthusiasts. Then three factors helped popularize the sport during the 1980s (Mckhann, 2001). First, materials and technology borrowed from ski manufacturing made it easier to ride on snow. For example, manufacturers, to facilitate turning, added metal edges and produced snowboards with narrower centres. Second, a skateboard revival in the 1980s helped popularize snowboarding, when skateboarders took it up as a winter alternative. Third, ski areas began to accept snowboarders. In 1983 less than 10 per cent of American ski areas allowed snowboarding, but by 1997 very few of them prohibited it.

Its popularity grew rapidly in the 1980s and 1990s, in part because it is relatively easy to learn. Most riders attain a degree of proficiency after only a few sessions of instruction. As the following paragraphs show, the sense of freedom the sport offers in equipment, technique and choice of terrain adds greatly to its appeal.

Snowboarding the backcountry has much the same appeal as backcountry skiing on powder slopes. A mother who taught her five-year-old son to snowboard said:

> There are so many reasons I love snowboarding – the quiet of the snow, the beauty of the mountains, just being AWAY from my cell phone, job, bills, and responsibilities, even for just a few hours – but the joy of watching The Boy jump and spin and race me down the hill is the newest and best reason.
>
> But of course, being me, I wear my heart-rate monitor when I board and count the calories I burn. It's still a form of exercise, even if it's incredibly fun!
>
> (http://homeboyski.com/2007/11/27/im-a-skier-and-i-love-snowboarding, retrieved 18 January 2009)

Mary O'Connor of MountainZone.com writes about her interview with veteran professional snowboarder Craig Kelly:

> In an industry where depth refers only to snowpack, the fact that Kelly talks about spirituality, connectedness, and being in tune with the mountains, makes him a standout. At the same time, he seems to have no interest in standing out. After spending years in the snowboarding limelight, he's had enough fanfare.

Though he holds seven World Championship titles, it's been nearly a decade since Kelly competed. These days, he concerns himself completely with backcountry riding, which, he is quick to point out, is merely a venue for riding – not a way to do it. As ski areas became more and more crowded in the late '80s, he and many riders like him were simply forced to look for the goods out of bounds.

'I've always been into powder – that's what I like, riding. And now you have to go further to get it. So I've been pushed out there further,' he says.

Turns out, that's not such a bad thing. Kelly earns most of his turns these days (at least those that aren't serviced by a SnowCat or helicopter), spending many hours enveloped in the great wide open. It has become a sort of spiritual playground for him. Backcountry became not just a place, but a state of mind.

'There's just a feeling you get from certain things you do in life that just kind of feel pure and independent of what's actually, physically, going on. All of a sudden you have this feeling of clarity. Backcountry snowboarding has really done a lot to boost that feeling in me,' he says.

Though he lives and breathes a snowboarding life, he's not quite in line with the culture of snowboarding ....

But I prefer it that way because what I do, my experience and what I do in snowboarding, is really pretty independent of snowboarding and the more independent it is, the more pure and better I feel about snowboarding.

That's not so surprising, considering all of the hype and the world of difference between the quiet solitude of an untouched high-alpine glade and the DJ'ed, bro-brah, name-tossing, go-big-or-go-home atmosphere of today's contests. The latter has a pretty short shelf life.

'I see a lot of attention coming to young riders that are really good but I hope, for their sake, and their love of snowboarding, that they'll get to go through the phase that I went through, of being intensely interested in progression; getting somewhere as a pro, and then finding a release for that.' he says. 'Because if you go too far with it, it's pretty hard to stay in love with the sport.'

So what's left for this veteran pro to do? Plenty. As long as Burton keeps making boards, he'll keep riding and helping design them; as long as it keeps snowing, he'll find new terrain to explore; as long as there's more to learn about himself and mountains, he'll learn it.

Elliot (2004) studied what many snowboarders call 'soul riding', which is their label for the 'spirituality' they feel while free riding. Whereas by no means all the elements of soul riding that he discovered bear on the present discussion, the following do. One is a feeling of risk and awareness of possible death, which the free rider tries to balance against the desire for adventure. Another is a sense of play, as expressed in the opportunity to learn through pain, stress and the experience of pure flow founded on mastering the physical rhythm of snowboarding. Freedom and escape constitute a third element, seen above in the personal agency of Kelly and the mother of the five-year old. It comes with backcountry snowboarding, particularly the lack of regulations and the range of options to ride anywhere on a slope. A fourth element is that of awareness of nature. 'There is something about gliding through untouched powder up to the waist, soaring over rugged, windswept cliffs and weaving through trees that snake down the mountain like a mysterious maze that is worth risking death' (Crawford, 2008). These sentiments are consonant with a backlash movement within the hobby against bureaucratization and regimentation that has accompanied its professionalization and movement towards intense international competitions such as the Olympic Games (Burton, 2003; Humphreys, 2003). The awe and wonder of snowboarding is, these writers maintain, seriously subverted by these trends.

Skijoring

Curtis Olson of the University of Minnesota defines *skijoring* (modification of Norwegian *skikjøring*, from *ski* + *kjøring* driving) as a 'Norwegian word that means "ski-driving" – that is, a cross-country skier utilizing a dog or dogs as draft animals' (http://www.me.umn.edu/~curt/Skijoring, retrieved 18 January 2008). A website developed by the Alaska Skijoring and Pulk Association (http://www.sleddog.org/skijor/start.html, retrieved 19 January 2008) states that almost any fully glide-waxed ski will work for this hobby. Skate skis work well on groomed trails or with fast dogs. One can use diagonal stride skis on narrow, un-groomed trails. Metal-edged skis are not recommended, because of possible serious injury to the dog. The skijor belt should be at least three inches wide, most of it being padded for comfort. Many belts have leg straps that go around the upper thigh to hold the belt in place. It is connected to the skijor lines by a 'quick release'.

The dog, which is trained as a sled dog, must also be equipped, the above-mentioned association recommending a sled dog racing harness.

Nonetheless a new design called the European Skijor Harness (Guard Harness) has just reached the market; it more evenly distributes the force of the dog's pull. The skijor line is also important. It should measure from 7 to 20 feet, depending on the number of dogs pulling the skier. This line also has, for benefit of both skijorer and dog, a bungee section (20–60 inches long) for absorbing the shock of starting, stopping and travelling over rough terrain. Almost any breed of dog can be trained for skijoring, but the animal must have sufficient size and weight to accomplish the task with ease.

Though dogs appear to be the most common draft animal for this hobby, horses, even snowmobiles have also been used. Skijoring was a demonstration sport in the 1928 Winter Olympic Games. 'Grassjoring' is a summer version of the winter prototype.

Skijoring is basically cross-country skiing, requiring the same skills and producing the same wonder-filled rewards of being in snowy winter nature. Note, however, that human agency is diluted in this hobby, because much of the forward movement is provided by the dog rather than the skier. Still this feature has an attraction of its own: engaging in the hobby with a beloved pet, which one website claims the pet adores. Furthermore pulled by a powerful, fit dog a skijorer may travel at significantly greater speeds for significantly longer distances than possible without such assistance.

CANINE ECSTASY

Dogs love to skijor. They enjoy the exercise, meeting other dogs, outdoor scents, occasional wildlife encountered and of course companionship with their owners. From a dog's perspective, skijoring is as much fun as a walk times ten. If your dog jumps up and down when the leash comes out, just wait until she discovers what the harness foretells!

(http://www.skijornow.com/skijornowhome.html,
retrieved 19 January 2008)

Most skijorers, the above website states, are hobbyists with little interest in racing. They prefer to skijor either on prepared trails or off-trail in forested terrain. Nonetheless championships are held annually for sprint and long-distance racing for those who like to spice up their natural challenge with some inter-human competition. The competitive skijorer's sense of fulfilment and awe of nature are similar to that of the Nordic ski racer.

A skijoring resort in Alaska, through its advertising, tells of some of the awe-inspiring qualities of the hobby:

> Welcome to skijoring in Trapper Creek, home of some of the finest snow conditions in Alaska. This area has long been a favorite spot for many winter sporting activities, and I strive to offer unique opportunities for novice and experienced skijoring enthusiasts.
>
> These outings will take you to an area with magnificent views of Mt. McKinley, across a landscape easily traversed by any skier.... Skijoring can be enjoyed by people of all ages. Adding a dog to your skiing adventure can allow you to travel greater distances in less time. I do, however, recommend you have some cross country skiing experience before you try skijoring.
>
> Come and enjoy one of the most invigorating of winter sports .I will accompany you on a pleasurable outing you won't forget, or set you up for your own adventure. Your experience can include a magical Alaskan full moon, or Northern Lights viewing, nighttime skijor, if the timing is right.
>
> (http://www.alaska.net/~pattyc/skijor.htm,
> retrieved 19 January 2008)

Dogsledding

Some of the popular literature on skijoring likens it to dogsledding. Their similarity lies strictly in the training and commands for the animals, for the skijorer is basically a skier and the 'musher' (driver of sled dogs) is primarily a rider. Butcher and Sassi (2007) say that dogsledding originated among the Inuit living in the Arctic regions of Alaska, Canada and Greenland. The first sleds were constructed with local natural materials and used for transportation. Today the basic dogsled is made of wood, metal, and plastic, however, and is mostly, though not exclusively, a recreational vehicle now enjoyed both within and outside these regions.

A dog team pulls the sled and its musher over snow and ice, while the latter stands on, pushes or runs with it. The skill and knowledge needed to form, drive, and care for a dog team – the serious leisure component of this hobby – are evident in the following passage:

> Mushers stand on the back of the sled and direct the dogs with voice commands. The most common are *gee* for right, *haw* for left, *hike* to go, and *whoa* to stop. Drivers make the sled run smoothly by shifting their body weight around turns, pedaling with one leg, and getting

off to push or pull the sled. A musher's most important responsibility, however, is to the dogs: feeding and watering them, checking their health, and tending to injuries.

To glide across snow and ice smoothly, the sled rides on two runners that extend several feet behind the main portion of the sled. The musher stands on these extensions while the sled is moving and holds on to a vertical piece of wood called a *handlebow*.... The driver can stop the sled by standing on the brake, which is a set of steel claws that drag in the snow. Once the sled is stopped, the musher can tie it to a post with a *snowhook* or a *snub line*.

Mushers must know basic veterinary skills to care for the team's health. While traveling, mushers closely monitor their dogs for fatigue and injuries. Most vulnerable are the paws, which suffer cuts and abrasions from the snow and ice. Mushers treat sore feet with balms and protect them with fabric booties....

Sled dogs are raised [typically by a musher] to enjoy running. Once they are old enough, they are placed on a team and tested in different positions [e.g., leader, middle, wheel]. Each dog then assumes a specific role within the team.

(Butcher & Sassi, 2007)

In short, hobbyist dogsled drivers need sufficient physical conditioning to occasionally run with or push the sled as well as to suddenly shift their weight to facilitate forward movement. Experience tells when best to engage in these core activities. They also need to learn how to raise, train and care for their dogs.

The wonder of nature felt while mushing a dogsled is similar to that felt in skijoring and cross-country skiing: skimming quietly over the snow in a hushed wintery setting that unfolds continuously during the journey. Drivers can always stop the sled and look around or take time for a rest. If not constrained by forest or geographic features, they may choose a route that interests them most. In other words mushers find scope for their own agency.

The competitive side of dogsledding dilutes these sensations, but adds new ones. Dogsled races fall into three categories. Sprint races are limited to distances that dogs can sprint or lope. Middle-distance races must be less than 500 km (300 miles), with long-distance events being over 500 km. Most race trails are groomed and lined with coloured markers. The terrain, like the weather, may be harsh and inhospitable. Snowdrifts, blizzards and thaws can, at almost any time with little notice, create difficult racing conditions. Drivers need knowledge and experience to successfully negotiate these contingencies.

The Iditarod Trail Sled Dog Race is the best-known annual competition in this hobby. The 1852 km (1151 mile) historic Alaskan trail saw its first dogsled race in 1972. Butcher & Sassi (2007) estimate that about 4000 dogsledding competitions are now held annually throughout the world. Local and regional organizations oversee these races, enforcing standards set by the global governing body, the International Federation of Sledding Sports.

Kicksledding

The contemporary model of kicksled was first used in middle and northern parts of Sweden in the late nineteenth century. At that time they were practical vehicles for transporting people and goods (http://www.lahdenmuseot.fi/asiakkaat/lahdenmuseot/www.lahdenmuseot.fi/content_images/asiakirjathimrtm/history_of_kicksled.pdf, retrieved 22 January 2008). Only recently have they caught on in North America, and here primarily as a means of leisure. The Nordic kicksled resembles a dogsled, but is shorter, lighter and requires no dog power. Sledders propel it on snow or ice by kicking back with one foot, while the other remains on one of the sled's runners, not unlike the way a child propels a scooter or a skateboard. For balance and steering, a handlebar is fixed to the runners, and a seat that can carry a passenger is mounted in front. Special shoes for kicking on ice are available.

Riders can kicksled almost anywhere there is snow or ice. Moreover, like those who push a scooter or walk, kicksledders come from nearly all age groups. The attractions in the natural environment of winter are much the same as those encountered while walking or skiing there. As described this activity is casual leisure, unless done in competition.

Competitive kicksledding requires the rider to be physically fit and capable of managing the sled at fast speeds uphill, downhill, around corners and the like. Here there is acquired skill and knowledge about how sleds work under these conditions, not to mention the conditions of the ice and snow on which they run. At this level kicksledding is hobbyist serious leisure. Championships are now held in Europe, although the largest of them – the Eurocup – does not always take place annually (http://www.iksaworld.com/welcome.html, retrieved 22 January 2008). One of the longest of the Eurocup races is run in Finland; it covers 100 km.

Snow kiting

Snow kiting is a winter board sport. It combines the airfoil and techniques used in kite surfing on water with the footgear and gliding surface used in snowboarding. In lieu of a snowboard some kite skiers use alpine skis or Telemark skis. In the early days of snow kiting, foil kites were the most common type, whereas today some kite surfers use their water gear for snow kiting. Tube kites are also used by some. Snow kiting differs from other alpine sports, in that, when the wind is blowing in the right direction, it is possible for the snow kiter to travel uphill with ease. Snow kiting is becoming increasingly popular in countries where skiing and snowboarding are also in vogue, among them Switzerland, Austria, France, Iceland, Norway, Sweden and the western United States.

According to Snowkiting.com kite skiing got started in the mid-1980s after some alpine skiers used a re-bridled square parachute to ski upwind on a frozen bay in Erie, Pennsylvania. After this

> kiteskiers began kiteskiing on many frozen lakes and fields in the US midwest and east coast. Lee Sedgwick and a group of kiteskiers in Erie, PA were the first known ice/snow kiteskiers. Snowboards were not used until many years after, as snowboards have a difficult time on ice and extremely hardpacked snow. Ted Dougherty began manufacturing 'foils' [an airfoil is a wing-shaped structure that gives rise to a lift force when moving through air] for kiteskiing and Steve Shapson of Force 10 Foils also began manufacturing foils using two handles to easily control the kite. In the mid 1980's Shapson, while icesailing, took out an old two line kite and tried to ski upwind on a local frozen lake in Wisconsin. Shapson introduced the sport of 'kiteskiing' to Poland, Germany, Switzerland and Finland. Shapson was the first person to use grass skis to kiteski on grassy fields. It's believed that some European or Swiss kiteskiers coined the word 'snowkiting.' The following terms describe the sport of 'Traction Kiting' or some refer to as 'Power Kiting': Kite buggying, kiteskiing, kitesurfing, kiteboarding.
>
> (http://snowkiting.com/node/191, retrieved 22 January 2008)

The size of kite depends on the terrain on which the skier will be riding (e.g., packed snow, powder snow) and wind speed. Other factors affecting movement include weight, strength, skill and experience of the skier.

Snow kiting is challenging, with lessons commonly sequenced from beginner through intermediate to advanced. The latter includes instruction in various 'tricks' such as Back Rolls, Front Rolls, Spins, Kite Loops and Front roll Kite Loops. The ideal place for snow kiting is a very large snowy field, about twice the size of a football field, with constant wind. Alternatively a large frozen lake may be used, preferably covered with snow. Here novices learn how to lay out the kite on the snow, grip the control bar, position themselves relative to the wind and start moving by wind power along the snow and, later, through the air. With skill and experience they get better at this, gaining greater control and becoming inclined to try more advanced manoeuvres. In the absence of snow on a lake, the hobby becomes ice kiting. Traction kiting is either ice or snow kiting in which the participant is pulled by a motorized vehicle.

The primary reward in a successful outing of snow kiting is effortlessly speeding with the wind over the snow and, for advanced hobbyists, successfully completing any of the aforementioned tricks. Unlike other downhill or alpine sports, snow kiting can be done uphill given the right wind direction – another prized experience. While in flight this hobby offers unadulterated flow to its participants. Additionally snow kiting is lauded for its freestyle approach to execution, indicating that human agency is given maximum expression here.

> Rachael Miller screams across Lake Champlain on skis, traveling close to 30 miles per hour while tethered to a kite that obeys her like a giant, if somewhat obstreperous, dance partner.
>
> The pair literally whistle in the wind as they skip across cloud shadows from the sun above the Adirondack Mountains. Splashes of powder from the modest overnight snowfall explode under Miller's skis.
>
> It seems her only limit is the boundary between ice and water, located somewhere far over the horizon. The scene is dreamlike...
>
> Time for a reality check.
>
> Here is a 35-year-old woman emitting occasional joyous yelps as she cuts new tracks through an endless expanse of untracked powder snow.
>
> She is going about twice the speed of the wind. And if she chooses, she can jump 30 feet high in what may be the only sport where you go up faster than you come down. There are no lifts, no lift lines, no lift tickets, and no out-of-control skiers and snowboarders threatening from above. In fact, there is no above; there's no hill – or hardly any gravity, so it seems.
>
> (Arnold, 2006)

Competitions are held in various parts of the world, centring primarily on excellence in freestyle (executing the various aerial manoeuvres) and racing. In 2008 there was, for example, the Snowkiting Austrian Open held on 2–3 February and the ISKA (International Snowkiting Assoc.) World Championship 2008 Kitesfera Challenge held on 15–17 February in Slovenia. There are also non-competitive festivals. They feature snow kiting lessons and the hobby in action, especially its spectacular tricks (e.g., Kitestorm, held annually in February in New England).

Snowmobiling

The snowmobile is a vehicle powered by a gasoline-engine and designed for rapid travel over snow-covered terrain. Traction generally is provided through a drive track, a belt of rubber reinforced by steel rods, at the rear of the vehicle. Snowmobiles usually cruise at 80 km/h (50 mph), but some racing models with more powerful engines can reach speeds up to 190 km/h (120 mph). The typical snowmobile is steered by handlebars attached to a pair of skis at the front of the body. Development of the snowmobile generally is credited to an American inventor, Carl J. Eliason, who constructed a motorized toboggan in 1927. About 95 per cent of the snowmobiles sold in the United States are used for winter leisure (*Microsoft Encarta Encyclopedia Standard* 2001, 2007a).

We focus here on the small machines – ones designed for a driver and possibly another adult rider – rather than the large multi-passenger vehicles. The latter are primarily used for work and commercial purposes. Most non-competitive snowmobiling is done on established trails at the level of casual leisure participation. People may rent a machine for a few hours from an outfitter, much as they would rent a moped. At the casual level loss of control is possible, sometimes caused by a failure to realize how powerful the machine is or by a driver losing his or her grip on the machine. *H-C* Travel, a tour operator in the United Kingdom, advertises that snowmobiling is 'very easy and requires no special skills or vehicle license. Speed is controlled by a thumb-activated throttle. Braking is done with one hand while steering takes place through a handle bar' (http://www.hctravel.com/html/snowmobiling.html, retrieved 23 January 2008).

Nevertheless snowmobile racing, depending on the kind of race, can be a nature challenge hobby. 'Snocross', which is done on a course similar to motocross, is serious leisure but not an NCA. The course is prepared. On the other hand, off-course races require skill and experience, especially in turning at high speeds; they constitute snowmobiling as an NCA. Consider oval racing on, for instance, a snow-covered lake with the oval marked off as a minimally prepared natural course (see next section on ice skating for the attractions of frozen lakes and rivers). The 'Iron Dog', the longest snowmobile race in the world, is held annually in Alaska. It runs 1971 miles starting at Wasilla then passing through Nome to Fairbanks. The name recognizes the long-standing popularity of mushing in Alaska, particularly the Iditarod race. There are also linear acceleration races, or drag races, for snowmobiles such as the Western Canadian Ice Race Championship now held annually in Alberta, Canada (http://www.classicwheels.org/wfs/wcric, retrieved 23 January 2008).

Finally, a third type of snowmobiling known as 'mountain climbing' might be classified an NCA were it not highly risky. This hobby, often the subject of dramatic videos, is accomplished with high-powered machines. The publicity for the video '2 Stroke Cold Smoke 7' shows at once the skill needed, the drama involved and the inherent danger in mountain climbing on a snowmobile:

> The latest snowmobile video from Frontier Films will blow you away. Backcountry, Backcountry and more Backcountry. Riding at its best. Huge cornice drops, gigantic step ups, upside down whips, record distance jumps, and pure adrenaline. 45 minutes of fast paced action with a rocking soundtrack. Plus our behind the scenes bonus film. Shot in Alaska, Utah, British Columbia, Yukon and Montana. Starring: Dan Gardiner, Ross Mercer, Dan Phillips, Randy Sherman, Ryan Britt, Tyler Nelson and many more. One of the top snowmobile videos out for sure!
>
> (http://www.extremesportsvideos.com/snomobil.html, retrieved 23 January 2008)

Another high-risk snowmobiling activity is 'high marking', defined as 'when a snowmobiler ascends a slope to the highest point they [sic] can reach. Also known as hill climbing' (from an online encyclopaedia developed by the National Avalanche Center and The Friends of the Utah Avalanche Forecast Center, http://www.avalanche.org/~uac/encyclopedia/high_marking.htm, retrieved 23 January 2008). This practice is also extremely dangerous, as the following list of tips from this online encyclopaedia attests:

Snowmachine High Marking Tips:

1) Start out on gentler slopes and work your way up to steeper slopes as so you can test the stability of the snow. Start on the side of a slope instead of center-punching it. Do your first runs low and fast rather than climbing as high as possible right away, which leaves you very committed and vulnerable. If possible, do your first runs on the more dangerous slopes from the top down to improve your chances of escape.

2) One at a time. If a person gets stuck, DO NOT SEND A SECOND SLEDDER TO HELP!!!! Fact: Roughly 33% of snowmobile fatalities occur when a sled is stuck. About 34% involve more than one machine on a slope at the time of the avalanche.

3) Everyone else should watch from a safe spot. Always park well away from the bottom of steep slopes or off to the side of the avalanche path. Don't count on being able to outrun a slide, but just in case, get in the habit of parking parallel, facing away from the avalanche path, rather than one behind the other and always leave the kill switch up when you shut down your machine.

4) Be wary of steep, smooth, leeward slopes. Slopes that have been stripped by wind (windward) are usually safer than slopes that have been loaded (leeward).

5) At the top of the high mark, turn towards your escape route instead of away from it and make your turn quickly, while you still have enough speed built up to avoid getting bogged down or stuck on the turn.

6) If unsure of the snow stability, favor slopes that have recently avalanched over those that have not yet slid. You can still sled on unstable days – just chose slopes less than about 30 degrees that are not connected to anything steeper. On some days the snow-pack is just too unstable to risk highmarking, and carving powder in the flats is the most prudent choice.

7) Avoid slopes with deadly terrain traps such as gullies, steep-sided creek bottoms, or slopes that end in depressions because of the high probability of a deep burial. Do not ride on slopes with cliffs below. Favor slopes that are fan-shaped at the bottom and do not have obstacles like rocks or trees to crash into. Concave bowls are nasty traps because the fracture propagates around the slope and all the debris collects at the bottom like a huge funnel, which could easily bury you under 10 to 30 feet of debris.

Ice

From the standpoint of NCAs on ice vis-à-vis on snow participants face some distinctive conditions. One, snow is commonly far more prevalent than bare ice, the required surface for all ice-based NCAs. This difference is compounded when ice that could be used for an NCA fails to form because ambient temperatures are too warm. Two, sometimes the needed bare surface is only available after snow is cleared from it, in some situations a laborious task. Three, ice is not as readily renewed as snow is. That is, a snowstorm brings a fresh surface, whereas to achieve this naturally with ice requires a period of thawing followed by one of freezing temperatures, a much less common occurrence during most winter months. Four, ice compared with snow is riskier, in that everywhere it can give way, often plunging the startled victim into extremely cold, deep water clad in heavy winter clothing serving as an anchor. Snow is risky only in avalanche zones. Five, spaces of ice are usually exposed to the wind, making wind chill a more routine consideration in most ice-based hobbies than those based on snow. The latter, if necessary, can usually be undertaken in or near protective stands of trees.

Because usable ice is generally harder to find than usable snow, people in search of the former generally have to travel farther to partake of their ice-based hobbies than those based on snow. Moreover, once there, the other four distinctive conditions may work adversely at any time to discourage participation. All this helps explain why the ice-based NCAs, though nearly as numerous as those pursued in snow, appear to interest many fewer people. This ratio is reversed, however, when considering all the ice-based serious leisure done in artificial settings such as municipal skating rinks and ice hockey facilities. Ice skating exemplifies well this distribution.

Ice skating

When done as a hobby this activity is purely serious leisure, in that people can only skate if they have the requisite skills, which require routine practice and conditioning. There is skill in being able to balance oneself on the thin steel blades fixed on skating boots and then to push oneself forward and backward and to turn and stop on them. Aerobic conditioning is needed to do this for any length of time, as well as muscular conditioning and strengthening of a range of lower-body muscles (especially those around the ankles). The activity, using animal bones tied to the feet, is estimated to date to 10,000 B.C. (*Microsoft Encarta*

Encyclopedia Standard 2001, 2007b), though the modern skate could only have been possible with the invention of metal.

Ice skating qualifies as an NCA only when done in a minimally modified outdoor setting. Minimal modification means, by and large, clearing snow off naturally formed ice to create a rink or trail. Such rinks and trails are established on lakes, ponds, canals, rivers and the like. Still, with luck, these natural features may have no snow to remove, because none has collected or the wind has cleared the surface. By these criteria flooded community ice rinks, say, in parks do not count as places for skating as an NCA.

Skaters experience the wonder of nature in these four settings in several ways. One is being on top of a frozen lake, river and so forth in the fresh air and moving along at an exhilarating speed under one's own steam and skill (agency). Moreover, on a river or canal (e.g., the canals in Holland and the Rideau Canal in Ottawa, Canada), the winter trees and vegetation greet the gliding skater in much the same fashion as they do the cross-country skier or kicksledder. Falling snow, providing it does not become an obstacle, adds to this natural wonder, as do interesting patterns of cloud, sun and shadows.

So far we have described the NCA of non-competitive general ice skating: skating for the pleasure and self-fulfilment gained from polishing and using its basic skills in an outdoor setting. *Figure skating* with its characteristic turns, spins and other movements builds on this foundation, and may be practised outside on natural surfaces. Some general skaters have also learned some of these manoeuvres, often doing them as part of an outing. The same may be said for those who go in for *speedskating.* These skaters, wearing special skates with long blades, find considerable fulfilment and exhilaration in zipping across the ice at noticeably greater speeds than the ordinary general skater

General skating is not competitive, whereas both figure skating and speedskating may be. Nowadays, figure skating competitions are held indoors, and as such are not NCAs. It is much the same for speedskating, which as competition is commonly conducted on prepared ovals, many of which are indoors. Those held outside take place in ovals constructed in an open-air stadium or park in circumstances too artificial to offer a nature challenge.

Motorized ice racing

While competitive speedskating and figure skating are almost exclusively conducted in artificial settings, motorized ice racing has taken to the slick surfaces of lakes and rivers. Such competitions are held for motorcyclists, car drivers and drivers of all-terrain vehicles (usually quad runners). There is no doubt about the considerable skill and experience needed to do well in these contests, much of both having to do with, for example, how to start and stop and how to control skidding on the slippery surface, which varies with temperatures and whether it has a dusting of snow. Studded tires help a great deal in this regard, so to boost the challenge of racing on ice, competitions are also organized in which only ordinary tires are permitted. Some outdoor races proceed around an oval, whereas others are linear acceleration events or contests over courses consisting of numerous turns and straight sections.

As with many other non-human means of locomotion, these types of ice racing dilute the sense of the ambient environment as drivers compete intensely with one another. At the same time the sense of speed in the outdoors is certainly enhanced through the use of powerful motorized vehicles. That drivers are confined to a track would seem to undermine somewhat the expression of human agency. Furthermore, as a protective measure, motorcycling ice racing calls for banks of snow or bales of straw along the perimeter of the race track. This amounts to a substantial modification of the natural setting. A car ice racer recounts his impressions as he drove a course filled with numerous turns:

> Safely past the spinner I flick my car into the left hander leading onto the front straight, upshift to third, and smoothly accelerate onto the straight in a nicely controlled third gear drift with a minimum of wheelspin.
>
> All the way down the straight you can hear the engine revs hunting up and down as the rear wheels scrabble for grip on the slick ice.
>
> The turn-in to the hairpin could have been cleaner. I make a small bobble trying to drag the car up against the inside snowbank where a narrow strip of better traction can still be found.
>
> On the back side of the track, turning into the new right/left kink, I've finally discovered that it's smarter to enter this first turn slowly in order to hook the inside on the exit, thereby setting up a much straighter run through the left turn onto the short back straight.
>
> Oh ya. That feels MUCH better.
>
> While busy congratulating myself for discovering this improved line through the new kink, I completely miss the upshift from 2nd to 3rd gear. Doh!
>
> The back 1, where I spun earlier, is actually a double apex turn.

For the first time I manage not to overshoot the first apex going into this turn. And I successfully use the heavy rear end momentum of this car to keep the tail swinging all the way around the double apex hairpin in one long continuous slide.

Listen to how little drama is coming from the throttle department.

My confidence level is starting to come back up again and I'm slowly, but steadily, catching Colin's car.

(http://icerace.net/Vette03/Prac1c.htm, retrieved 24 January 2008)

Unlike snowmobiling these NCAs are dominantly, if not entirely, competitive. People seem not to drive cars, motorcycles and quad runners on ice purely for the natural challenge it offers, for such activity is not, in fact, much of a challenge.

Ice boating

In this winter sport, boats with large sails are propelled by wind over frozen bodies of water on sharpened runners. Sailor Randy Rogoski says that modern ice boating (also called ice yachting) began in the mid-1600s in Europe (http://www.iceboat.org/ice, retrieved 24 January 2008). The original ice boats were sailboats with a strong cross plank under the hull near the bow and fitted with runners. They were used to move cargo on the frozen canals of The Netherlands. Premodern ice boating originated more than 4000 years ago in northern Europe (*Microsoft Encarta Encyclopedia Standard* 2001, 2007c). The first ice-boat runners at that time may have been made from bone. Iceboats were introduced to the United States during the American Revolution (1775–1783).

The first organizational body for ice boating, the Europäische Eissegel Union, was established just prior to the First World War. This group held several annual international ice boating championships. By the early part of the nineteenth century, ice boating had become popular in the United States, even though most of the boats manufactured were large and expensive. But around 1930 smaller boats began to appear, stimulating a rapid growth of interest in the sport (*Microsoft Encarta Encyclopedia Standard* 2001, 2007c).

The DN-60 of today, the most commonly used ice boat the world over, has a medium-sized (12-foot) hull in which the skipper rides. The hull is crossed with a runner plank about eight-feet long, and the entire structure is powered by a 60-square-foot sail (seemingly large for so small a boat). The DN is economical to build or buy and can be transported on a car roof.

The skill, knowledge and experience of ice boating are all evident in the following account of sailing on Lake St. Clair outside Detroit:

> Winter sailors zip along frozen waters at speeds of 50 to 100 miles per hour, a feat made possible by aerodynamics that accelerate the craft to three times the speed of the wind force. Though sailed like a boat in water, ice boats are trickier to handle when it comes to turning and stopping because of the speed and the lack of friction of steel blades on ice. In heavy winds, they often hang on the edge of becoming airborne. One hearty sailor put it this way: 'An ice boat is a beast once it gets going. The closing rates between ice boats and other objects are so fast that if you have to think about what to do, it's probably too late.'
>
> Sailing expertise, cold weather, speed and a sense of danger are all part of the appeal. Enthusiasts come from all walks of life, bound together by the sense that the rugged Viking-like sport has purpose in a world filled with creature comforts.
>
> (http://info.detnews.com/history/story/index.cfm,
> retrieved 24 January 2008)

Or consider this publicity from a sporting equipment dealer in Montana:

> Cutting across and [sic] icy expanse on steel runners with the wind in your sails means exhilaration to those who have discovered Ice Boating. Visit our local ice boating paradise on Canyon Ferry Lake below the Silos and you will soon see the fascination.
>
> Speed is not the only thrill. This is a sport which offers sailors the opportunity to enjoy the winter as well as the summer on the lake. All the skills of sailing fine boats come to play when you slip into the sleek hull of an ice boat. The perfect winds on Canyon Ferry Lake will take ice boaters anywhere they want to go. Zipping past frozen crystals on ice, over the glossy frozen wide open spaces can take your breath away.
>
> Add the blue sky and sunshine to crisp winter air and you will quickly discover why the shores of the boat launch are filed [sic] with ice boaters when the lake is frozen. Join Montana Kite Sports to discover this thrilling sport.
>
> (http://www.montanakitesports.com/ice,
> retrieved 24 January 2008)

For those who want to race in addition to pursue non-competitive ice boating, competitions abound. Each winter the International North American DN Ice Yacht Racing Association (IDNIYRA) holds its North American Championship at various locations in the northern American states and the Canadian provinces. The IDNIYRA Europe has held European Cup and World Cup championships for many years. As in sail boat racing on water, ice boat races are typically run on windward-leeward linear courses. In this way the boats usually sail across the wind.

Conclusions

This chapter demonstrates that harsh winters need not necessarily drive people to a warm climate for their duration. Interesting, fulfilling, awe-inspiring activities in the outdoors are available in considerable variety, albeit always conducted on either ice or snow. Moreover, spending much or all of a snowy winter where it is warm is beyond the means of many people, should they even want to do this. Their choice is either to find their leisure inside during those months or come to grips with how to find it out of doors. In this sense alone NCAs on ice and snow, by offering some cheaper local alternatives, help discourage pursuit of some expensive non-local interests.

Coming to grips with winter outdoor activities is, in fact, not that difficult. None of the activities covered in this chapter is inordinately expensive, with snowshoeing and ice skating possibly being the cheapest. Perhaps the greatest expense lies in clothing; winter hats, coats, boots, mittens and so on, if effective and durable, and especially if fashionable, are generally dearer than the leisure clothing of summer. And, as mentioned, participants often have to travel significant distances to pursue their winter hobby, a cost in both time and money. This said, these activities are accessible to a wide range of incomes. Moreover for those concerned about creating and living in an environment that is sustainable, these winter NCAs offer many possibilities. This leisure, unlike winter activities that require, say, constructing, operating and maintaining a building (ice hockey arena, indoor basketball court, indoor swimming pool), leaves nature unmolested. In other words, with the exception of snowmobiling (see Chapter 8), these NCAs may all be classified as environmentally sustainable undertakings.

Turning to gender, it seems not to be an issue, in that both sexes are well represented in them, even while historically women were under-represented, if not entirely absent, in such pursuits as snowmobiling, ice boating, dog sledding and snowboarding. And women may still be underrepresented in back-country snowboarding. Some of these activities lend themselves well to familial outings, including cross-country skiing, snowshoeing, ice skating and snowmobiling. Most are for the younger ages, but some elderly people do skate, cross-country ski or snowshoe.

We will close this chapter with some speculative observations on the class basis of these winter hobbies. The observations centre on taste for them rather than on their cost, already held not to be prohibitive for much of the population pursuing them. One, snowmobiling, dog sledding and ice racing appear to attract mostly working-class participation. Two, backcountry snowboarding and snow and ice kiting seem to be a young person's game. Both are relatively new, however, and we must acknowledge the possibility that they may remain popular with present-day enthusiasts as they age. Three, cross-country skiing, snowshoeing, ice skating and ice boating appear to hold greatest appeal for the middle and upper classes. As for skijoring and backcountry skiing, we lack the information needed to estimate their appeal according to class.

7
Mixed Activities

While our classification of NCAs by separate elements has served as a useful conceptual and organizational tool, it has been evident from discussions in the preceding chapters that a number of NCAs do not sit entirely comfortably in a single category. Indeed certain activities can be considered *hybrid* NCAs as they involve simultaneous engagement with more than one natural element. These include the wind-propelled activities undertaken on water, land, ice or snow. Other examples are caving, canyoning and coasteering, where participants may encounter both land and water within the bounds of a single activity. Mountaineers may also be challenged by a mix of rock, snow and ice, which they negotiate using a variety of techniques encompassed by one NCA. The combination of elements in hybrid NCAs can be part of their specific appeal. As the kitesurfer put it in Chapter 3, in these activities participants can feel that the 'boundary' between particular elements – in his case water and air – 'dissolves in a fluid transition'.

It is also often the case that separate NCAs are combined in a single outing, allowing participants to enjoy the range of challenges that this facilitates. For example, activities such as flying, hiking, off-roading, mountain biking, paddling, boating, sailing and cross-country skiing can be enjoyed for their own sake at the same time as they provide access to natural settings in which other challenges – fishing, hunting, diving, nature photography, rock climbing and so on – can be found. Wilderness journeys, particularly multi-day trips, may also necessitate a combination of NCAs depending on the terrain that is traversed. Canoe trips, for example, may require portages to link together bodies of water, or on hikes in mountainous areas a portion of glacier travel or the crossing of a snow-covered alpine pass may be needed. Participants are attracted to these types of journeys for the challenges demanded by the diverse terrain, as well as the sense of remoteness, self-sufficiency and exploration that such travel allows.

In recent years, the emergence of multisport and adventure racing have added competition into this mix. Indeed, undertaking a 'wilderness journey with a deadline' (Bell, 2003, p. 220) has become such a well-established pursuit in its own right that we considered it worthy of separate discussion as a mixed NCA. In this chapter we also take a look at a distinctive set of NCAs, spanning the full range of natural elements, in which the core activities are pursued primarily as volunteer activities. They are search and rescue, conservation work, activity-setting maintenance and education.

Multisport and adventure racing

Races combining various disciplines have been around since the early twentieth century. Modern triathlons, consisting of a non-stop three-part event involving open water swimming, road cycling and running, were first held in the United States in the early 1970s. As discussed in Chapter 3, in a classic triathlon, and a long-distance triathlon (also known as an Ironman), only the swimming leg qualifies as an NCA. More recent variations of the triathlon, which the International Triathlon Union (ITU) includes within its multisport category, can be considered NCA events. These are the winter triathlon (a run, mountain bike and cross-country ski, all on snow) and the cross triathlon (also known as an off-road triathlon consisting of a swim, mountain bike, and trail run).

Multisport events, combining two or more NCAs, appeared soon after the first triathlons. Two seminal races, the Alpine Ironman (skiing, mountain running and whitewater kayaking) and the Coast to Coast (a two-day event featuring road cycling, mountain running and kayaking), were held in New Zealand in the early 1980s. Around the same time, the first Alaska Mountain Wilderness Classic took place. In 1989, the Frenchman Gerard Fusil organized the Raid Gauloises, the first multi-day adventure race with mixed-gender teams of five, in New Zealand's Southern Alps. From there, adventure racing spread to other parts of the world, becoming firmly established in North America following the first Eco-Challenge race held in Utah in 1995.

Over the following decade or so, adventure racing rapidly evolved into four basic categories: sprint races, which are usually completed in 3–10 hours; weekend races, undertaken in 1–2 days; stage races, which can extend over several days, with competitors completing a different 'stage' each day; and expedition-style races, modelled on the original Raid Gauloises, that is, 'a non-stop, self-sufficient, multi-day, multi-discipline, mixed-gender team endurance competition that takes place in the wilderness over a designated but unmarked course' (Kay & Laberge, 2004, p. 154; D. Mann & Schaad, 2001).

The combination of disciplines varies from race to race, depending on the nature of the terrain in the race location and the inventiveness of the race organizers. The 'core' disciplines include foot travel (trekking, trail or mountain running), paddling (either rafting, canoeing or sea kayaking), mountain biking and basic mountaineering (including the use of rope techniques involved in rock climbing, abseiling, rope traverses and glacier travel). Winter races can consist of cross-country skiing, snowshoeing and other snow and ice skills such as glissading, ice climbing and digging snow caves for shelter. Other disciplines include horseback (and even camel) riding, canyoning, coasteering, open water swimming, river boarding and spelunking. Inline skating, while not an NCA, is sometimes used to cover sections of paved roads where they cannot be easily avoided. In most cases a sprint race combines a mountain biking, trail running and water leg – which could be an open water swim, tube float, raft trip or flat water canoe (Marais & de Speville, 2004).

Teams usually consist of between two and five members, while some races, particularly multisport and cross triathlon events, may allow, or be exclusively for, solo competitors. Typically, there will be transition areas where participants make the change from one discipline to the next, and where they can replenish food and water, change clothing and collect equipment needed for the next stage. In some races, teams are allowed their own support crews, while for others they must organize all their own supplies and equipment, which is then transported to the transition areas by the race organizers.

The natural challenge of multisport and adventure racing is provided by the combination of the various NCAs involved and – depending on the nature of the event – orienteering-like navigational challenges. In addition to having basic proficiency in a wide range of activities, adventure racers need the strength and endurance to complete the course for which they have entered. Longer team events, in particular, require varying degrees of self-sufficiency, strategic thinking and 'superhuman endurance' (Marais & de Speville, 2004). With the exception of some multisport events which take place on a marked course, navigation skills are a critical, if not deciding, factor in a successful race (Marais & de Speville, 2004). The course for an adventure race is not revealed until a pre-race briefing, when the competitors receive their maps and coordinates. Using map and compass (a GPS is usually banned) they must then plot their route and locate a series of compulsory control points that they must check into along the way. Races can involve moving in conditions of low visibility across challenging country, as well as travelling at night during multi-day races, and navigational errors can cost teams significant amounts of time.

As well as deciding their route, teams must negotiate their own pace through the race terrain, which could be anything from desert or rainforest, to alpine terrain and sub-zero temperatures. Pre-race preparation usually involves familiarization with the weather patterns at the race location and other aspects of local conditions such as tides, currents, times for sunrise and sunset and even moon phases, as a full moon on a clear night will aid night travel (Marais & de Speville, 2004). So while teams will know the general location and distance of a course and roughly how long it should take, they do not know in advance exactly the topography and vegetation they will be required to traverse, nor precisely the types of natural challenges they will face, and therefore whether their preparation has been adequate (Bell, 2003).

In coping with the unpredictable nature of an adventure race, teamwork is another factor that is critical for success. For a team to officially complete a race, each member must make it to the finish. Teams must be able to communicate and work together in an efficient and coordinated manner, drawing on individual strengths, whether they be endurance, problem-solving or technical skill, and accommodating and supporting each others' weaknesses (Edmonds et al., 2009; Marais & de Speville, 2004). The usual requirement for teams to be of mixed-gender can have further implications for group dynamics (Kay & Laberge, 2004).

On non-stop multi-day races, sleep deprivation is another significant challenge, with some adventure racers reporting levels of exhaustion such that they experience hallucinations (Bell, 2003). Jim Cotter describes the impact of fatigue on his team during the 1996 Eco-Challenge in British Columbia, Canada:

> Everyone goes through their own ups and downs, due to both energy levels and sleepiness. Generally, sleepiness seems to become worse at normal bedtime, and reemerges with a vengeance during the three to four hours before dawn. The low-energy (blood glucose) periods, sometimes referred to as 'hitting the wall,' are more frequent and unpredictable, though also more easily prevented or abated. Usually

at least one person is feeling OK, and their enthusiasm is generally a bonus rather than a pain in the arse.

(Cotter, 2003, p. 211)

Cotter's account of an earlier stage of the race highlights the challenges and rewards of adventure racing, as well as his appreciation of the beauty of the terrain through which they are travelling. This is something organizers take into consideration when selecting race locations and it is not uncommon for racers to talk about feeling that they are 'part of nature' (D. Mann & Schaad, 2001; Marais & de Speville, 2004). Cotter's team has missed a walking trail that was not marked on the map, resulting in an unnecessary struggle through thick vegetation:

The race is only one day old, and already we're shattered: physically from battling through the slide alder at a snail's pace after the morning run/ride, and mentally, by the realization that our valuable energy expenditure of day one has largely been in vain. After dragging out the sleeping bags and scoring nearly three hours sleep, we're revitalized and keen to make up some ground. The remainder of this first mountain stage leaves us no more tired but with several great memories. The fatiguing process of the slide alder and several more thousand feet of climbing valley walls and snowfields is quickly forgotten. Even the cold wind and discomfort of tying to sleep for an hour on a rock ledge above a glacier (4 am on the second night) is remembered less clearly than the view of an orange moon sliding up the side of a peak silhouetted above the blackness of a valley far below. This stage is otherwise highlighted by chatting to friends and friendly individuals from other teams whom we occasionally mingle with, by long periods of roped team travel across snowfields encircled by peaks, by mountain lakes and lush meadows in full bloom, and by beating the other three teams in our proximity out of the bush and down the forestry road to the trek-to-canoe transition area. Not that we're competitive!

(Cotter, 2003, pp. 209–210)

The number of adventure races worldwide grew throughout the 1990s. For the month of June 2010, more than 70 events were listed on one adventure racing website, in locations as diverse as Ukraine, Brazil, China, Ireland and Australia (see www.sleepmonsters.com/calendar. php). Adventure Racing World Championship events, organized and run by a New Zealand-based company, have been held annually since

2003. In 2009 the World Championship took place in Portugal, with 59 teams of four from 25 countries competing over a 700/800 km course for around 5 days or approximately 120 hours non-stop. An Adventure Racing World Series serves as qualifying races for the World Championship. In 2010 this series consisted of ten races taking place in locations across Europe, the Americas and Australasia.

Adventure racing participation rates in the United States have experienced a 50 per cent increase since 2006, reaching almost 1.1 million participants in 2009. So while overall participation is relatively low, it is one of the fastest growing outdoor activities in that country, along with off-road triathlons, for which participation has more than doubled over the same period, totalling 666,000 in 2009 (Outdoor Foundation, 2009a, 2010). Supporting and contributing to this growth are the adventure racing associations and clubs that are being established around the world to promote the sport. The Associação Portuguesa de Corridas de Aventura (Portuguese Adventure Racing Association), for example, has more than 400 athletes among its regular and affiliated members, and organizes the Portuguese Adventure Racing National Series (www.corridasdeaventura. pt/en/apca.php, retrieved 27 June 2010). Southern California Adventure Racing Buddies (SCARAB) has over 2900 members, and promotes itself as 'the largest adventure racing club in the world' (http://scarabs. homestead.com/, retrieved 27 June 2010). SCARAB provides opportunities for adventure racers to meet teammates, train together and exchange the latest techniques and event schedules. Websites like www. SleepMonsters.com – the name alluding to the battle with sleep deprivation experienced in multi-day races – provide news, event calendars, training tips, race reports and racer profiles. Training clinics and skills certification offered by adventure centres and companies are another means for participants to progress in this NCA.

Volunteering

In our final category we have a group of pastimes which combine the core activities of an NCA alongside various tasks within the context of career volunteering. As such they incorporate the characteristics of volunteering as serious leisure discussed in Chapter 1. In addition to their intrinsically satisfying nature, these activities have the appeal of helping to facilitate the enjoyment of and safe participation in the NCAs with which they are associated and accordingly may be motivated by either of the six types of interest proposed earlier; that is, (1) people, (2) ideas, (3) things, (4) flora, (5) fauna, or (6) the natural environment. These activities are search and rescue (SAR), conservation work, activity-setting maintenance and education. SAR occurs on land, water, snow, ice and in air-related settings, while conservation, maintenance and education can cover the full range of activity settings.

SAR as an NCA relates to operations undertaken to locate and recover missing and/or injured persons. SAR volunteers are typically persons adept in the relevant NCA, be it mountaineering, caving, canoeing, swimming and so on, who have acquired the additional skills and experience required to evacuate people from accident sites or extreme weather events for which regular emergency services are ill-equipped.

Lois (2004, pp. 121–122) describes the kinds of 'missions' performed by members of a volunteer SAR group in the Rocky Mountains in the United States.

> At times, members were asked to perform only slightly demanding, low urgency tasks such as hiking a short distance up a trail to carry a hiker with a twisted ankle out to the parking lot. Other times they were asked to perform very difficult, dangerous, or gruesome tasks such as rappelling down a 200-foot rock face to recover the body of a fallen hiker, traversing known avalanche terrain to reach a hypothermic snowshoer, and bushwhacking for fifteen hours over miles of treacherous terrain in search of a lost hunter.

In her 6-year ethnographic study of this group, Lois (2004) observed that in addition to technical skills such as competence on the terrain, outdoor first aid, communications, management of rope systems and other rescue equipment, SAR volunteers also had to develop skills to cope with the strong emotions that arise from demanding missions. Primary among these were feelings of fear, urgency and emotional upset at handling badly injured, distressed or deceased victims. Keeping these emotions in check was seen as important in order to perform necessary tasks competently. The satisfaction for participants came from successful missions, when victims were found safely and reunited with their families.

A SAR volunteer reflects on the satisfaction of finding a lost hiker.

> I still feel good about that. I mean it's [been] eight years, and I still feel good that at some point I made a difference in one person's life. If you can make a difference in one person's life that should be enough reward.

> (Lois, 2004, p. 146)

Helicopters have greatly facilitated SAR in mountain regions. Specially trained dogs and handlers are also used to help locate people buried in avalanches and in thickly vegetated areas where victims cannot be spotted from the air. The techniques used for cave rescues, which can be very slow and delicate operations, are highly specialized and adapted specifically to the confined nature of the underground setting. Among the water-based SAR activities, surf life saving is a well established volunteer group, particularly in Australia and New Zealand, as discussed in Chapter 3.

Conservation work includes activities such as invasive species eradication, habitat restoration (e.g., tree planting, dune restoration and clean water projects), threatened species monitoring and recovery, interpretation and advocacy. It springs from an interest in the natural environment, often nurtured by the special affinity an NCA participant feels for their preferred setting and the flora and fauna that inhabits it. An interest in preserving the integrity of the environment for future enjoyment may also be a motivator here, as we saw with the advocacy activities undertaken by groups like Surfers Against Sewerage (see Chapter 3).

Catherine Turner, a conservation volunteer in a rainforest reserve in Peru, describes her experiences which combine 'work' with the more leisurely enjoyment of nature and its challenges, including hiking, wilderness camping and bird-watching.

> ...camping out in the jungle, lying on my back listening to constant background noise of insects, frogs, birds and animals, watching the sky darken into night, watching stars appear above the canopy, watching the fireflies flit about, gleaming like stars about my head. It's an experience I won't forget, but we were there to work as well.

> The previous week had been spent setting up Mist Nets at intervals along the trail.... The fine, black mesh netting was set up to catch small birds that fly through the lower part of the forest as these are less easily monitored by our daily observations at the platforms set up around the reserve. Our job was to patrol the nets at hourly intervals and untangle any of the birds we found, take them back to the camp and weigh, measure and photograph them, before releasing them again....

> Through daily observations of the birds from platforms situated around the reserve, I was soon able to impress newer volunteers with my astounding knowledge of birds: Yellow-Rumped Caciques,

Russet-Backed Oropendolas, Vermillion Flycatchers, Greater Anis, Paradise Tanagers and the difference between the assorted parrots and parakeets that we spotted flying or perching all around us. Frog-hunts on wet evenings brought us the delights of amphibians; day-walks and night-walks enabled us to encounter groups of Squirrel Monkeys and Capuchins springing through the trees, a small black scorpion injecting its venom into a unlucky centipede, a night monkey with its big eyes shining back at us, and a rare sight of Kinkajou up in the branches.

(http://www.projects-abroad.co.uk/our-volunteers/volunteer-stories/?content=conservation-and-environment/peru/catherine-turner/, retrieved 17 January 2009)

Activity-setting maintenance and education includes activities such as building and maintaining hiking, mountain biking or cross-country ski trails, waste removal, guiding and instruction. These volunteer activities are focussed more specifically on facilitating participation in the relevant NCA and/or mitigating its impacts on the natural environment. Here the core activities are combined with additional skills from knowledge of the effects of erosion and the effective construction of hardwearing trails to the transfer of specialized knowledge to novices.

Conclusion

Concerns have been raised about the environmental sustainability of adventure races which can involve large groups of people traversing fragile landscapes in a context which may encourage speed over concerns for minimizing impact. However, many groups and individual adventure racers espouse environmental ethics. The US Adventure Racing Association (USARA), for example, has a code of ethics that includes practising minimum impact travel, not littering and respecting the land and inhabitants in race locations (http://www.usara.com/home.aspx, retrieved 25 June 2010). Races will typically have a set of guidelines to ensure environmentally sensitive behaviour with penalties for non-compliance by participants. Low-impact regulations include sticking to the designated trail to minimize erosion, avoiding damage to vegetation, carrying out all waste (including human waste in particularly sensitive environments), not disturbing local wildlife and using biodegradable soap (Marais & de Speville, 2004). Some of the larger expedition races also organize associated projects or donations to support local communities. The Eco-Challenge, in particular, claimed a high level of ecological and social sustainability. However, questions have been raised about the effectiveness of their environmental practices and their community projects have been seen by some as 'token gestures' (Bell, 2003).

As for consumption, like many NCAs, at an entry level adventure racing is not overly expensive, but costs can increase considerably with deeper involvement. Participants may transition into adventure racing from other NCAs, in which case they may already have much of the required gear. For first-timers, entering a sprint race is a recommended place to begin. In the United States a one-day event at a regional competition can cost around 100 to 200 US dollars per person and for these types of events most of the specialized equipment is usually provided and as the race is in a relatively controlled environment, little more than the correct clothing, footwear, bicycle, helmet and backpack is required. Longer races have more extensive 'mandatory equipment lists', which can include food, sleeping bags and/or some form of shelter, first aid and survival gear (Marais & de Speville, 2004, p. 23). Registration fees for longer expedition-style races may be several thousand dollars per team. In addition to the cost of the necessary equipment and travel to the race location, this is likely to be a barrier to participation. Kay & Laberge (2002) found that 61 per cent of those entering the Eco-Challenge expedition races were management-level corporate participants, and they explore the possible implications of this, including the parallels between corporate culture and adventure racing discourse.

Volunteer NCAs may be considered the most sustainable of all, particularly those with a specific focus on positive environmental results. SAR operations could be undertaken at some environmental cost, but typically only to the extent required to ensure the safety of those in distress. Given its public service nature, the cost of SAR equipment, training and activities is often covered by public funds and/or non-governmental organizations, corporate sponsorship and grassroots fundraising. Other volunteer activities may have access to similar funding, and if not, individual participants meet these expenses as part of their commitment to the activity and the natural setting in question. Such consumption is therefore focused on not only personal satisfaction, but also the achievement of desirable environmental and social outcomes.

8
Conclusion

In this chapter we address ourselves to the place of NCAs in society. In particular we consider their role in promoting fitness and their draw as tourist attractions. We also discuss how a number of these activities can engender personal tranquillity, based on the premise that nature often has a calming effect. In other words we look again at the psychological dimension of being at one with nature. Here we also weigh in on the role of consumption in experiencing nature through the pursuit of an NCA, noting where some of this leisure is virtually, if not literally, non-consumptive. That is, to find enjoyment and fulfilment in the outdoors and hence in life, it is not always necessary to buy facilitative goods and services. Finally, compared with the totality of casual leisure activities, a much larger proportion of the NCAs is kind to the environment. But only serious leisure NCAs can make this claim, since other outdoor serious leisure activities, like those of casual leisure, are likely to threaten sustainability (e.g., golf and golf courses, alpine skiing and ski hills, dirt bike riding and bike trails, and hunting and fishing when not regulated for sustainability).

All this is approached under the heading of the implications suggested by our findings, namely, those bearing on theory, policy, consumption and the environment.

Theoretic implications

The concept of the NCA is a useful addition to leisure studies theory, in general, and the serious leisure perspective, in particular. It is similar to one of its precursors, the outdoor activity, which refers to what people do for leisure in the open air. But outdoor activities may be pursued inside or outside cities and towns in natural or artificial

189

settings. Furthermore, contrary to the idea of outdoor activity, that of NCA includes a nature appreciation component (awe, wonder) while being restricted to activities that set a natural challenge to be met. Furthermore, with NCAs, we are able to distinguish this kind of serious leisure involvement out of doors from casual and project-based leisure pursued there. A casual stroll along a creek in the country or a helicopter ride to a mountain summit where riders then walk about offer very different experiences compared with kayaking the creek or climbing the mountain. Yet all four are commonly classified as outdoor activities.

Additionally the concept of the NCA when joined with that of flow enables us to understand more clearly than heretofore the nature of social high-risk ('extreme') sport and hobbies. Pursuing the core activities, many of the NCAs generates a significant sense of flow. Examples include white-water kayaking, cross-country skiing, backcountry snowboarding and wind surfing. Moreover being in flow in these NCAs, as opposed to experiencing leisure there more generally, presupposes *manageable* challenge, nothing too easy, which is boring, but nothing too difficult, which in some NCAs can be terrifying (another meaning of awe). Flow is most intense when people in their serious leisure operate at or near their mental and physical limits, but stop short of going beyond them.

By contrast, in *social high risk*, peers sometimes pressure individual participants in their company to engage in the activity at a level the second feel is much too risky; they see this as provocation to go substantially beyond their acquired skills, knowledge and experience. Here the challenge nature presents has become unmanageable and consequently flow either evaporates or fails to develop. This manageability criterion also applies to those who intentionally take dramatic risks for the fame and perhaps even the fortune that they bring. To the extent they are participating at a truly high level of risk – beyond effective control – flow cannot be experienced. It is likewise for people driven to satisfy a strong desire to experience an adrenaline rush in an activity, which can only come while being at risk. Risk for fame and fortune and for a surge in adrenaline are two additional forms of social high-risk.

Another concept comprising our theoretic list is that of adventure. Drawing on Johnson's (2003) definition we pointed out that an adventure need be neither dangerous nor occur in a challenging physical environment. In other words not all adventures are risky and some adventures spring from meeting challenges that are artificial such as those built into indoor climbing walls, velodromes and archery ranges.

Thus this book has further clarified the relationship of high-risk activity and adventure as these two bear on the NCA.

Implications for policy

In this section we cover two important implications for policy which follow from this study: fitness in nature and NCAs and tourism. This study also has implications for sustaining the environment, which however, is a policy matter of complexity great enough to justify separate coverage in a later section.

Fitness in nature

When it comes to generating physical fitness, the NCAs are by no means all of a kind. Some, like mountain hiking, cross-country skiing and multisport, are excellent aerobic conditioners, whereas others have the opposite effect, among them, skijoring, fishing from a boat and dog-sledding. Some require significant upper-body strength and are therefore good activities for building that capacity, for example, kayaking, canoeing, mountain climbing and long-distance swimming. Then there are NCAs that require considerable balance, including surfing, snowboarding, skiing (cross-country, water), ice skating and mountain biking. Yet, beyond this list, a good many NCAs contribute little or nothing to the fitness of their participants. Scuba diving, amateur astronomy, motorized ice racing and 4WD make up part of this category.

The wonders of nature may help pull some people towards one or more of these forms of fitness, providing a kind of motivation never seen in the use of artificial settings in towns and cities. But many of these wonders can also be experienced in sedentary NCAs, suggesting that the first will not usually be sufficient to coax people to engage in one or more fitness generating outdoor activities. Indeed it is quite possible that some participants pursue these latter activities with little or no intention of getting fit in the ways made possible by them. Rather they fall in love with the core activity of the pursuit, do it routinely and get in shape as an unintended result. A non-competitive interest in the activity and a passion for being in nature are crucial antecedents to this indirect approach to NCA fitness.

Whereas the NCAs as a group have a mixed record for enabling physical fitness, they fare much better in spawning mental fitness. In each chapter we explored systematically the proposition that each NCA examined there is pursued in its own awe-inspiring natural environment. Being in awe, or wonder, of an aesthetically attractive feature of nature, we have argued, is a major part of the psychological dimension of pursuing an NCA. Nevertheless Borrie and Roggenbuck's (2001) study revealed that the intensity of environmental awe may be variable, depending on what participants are doing, how long they have been doing it, meteorological conditions and so on. Nonetheless all NCAs at certain points in their execution engender significant tranquillity, which when coupled with wonder, helps relax the participant. This, in a nutshell, is the source of the mental fitness experienced while pursuing the nature challenge activities.

The NCA and tourism

A sizeable proportion of the NCAs nurture in their participants a desire to venture beyond their local opportunities for engaging in them, be that elsewhere in the region of the country in which they live, to another region of that country, or to a foreign country or region of the world. The world of tourist mountain climbing offers an example:

> A distinctive group of people around the world with sufficient money and time for going to destinations away from home travel specifically to climb one or more distant mountains. Not all these people are [NCA] hobbyists, however, since it is possible to climb some mountains as tourist objectives without the conditioning and knowledge needed in their serious leisure counterparts. Thus, Mount Fuji in Japan, which receives more than 200,000 visitors annually (http://www.mt-fuji.co.jp/info/info.html, retrieved 9 February 2008), has among other trails to the top, a relatively easy one for beginners, for casual leisure participants.
>
> (Stebbins, 2009c)

An elite branch of this hobby is comprised of mountain climbers who have travelled around the world to scale the highest mountain on each of the seven continents, collectively known in mountaineering circles as the 'seven summiteers'. They have conquered the 'seven summits'.

Although by no means all participants who want to follow their leisure passion elsewhere become NCA tourists (because, for example, they lack needed time or money), literally every NCA covered in this book can be pursued somewhere outside the local area where the participant lives. In fact some participants – big game hunters, oceanic scuba divers, white-water kayakers and canoeists and those who go in for multi-day canoe and backpacking trips, among others – are rarely able to enjoy their hobby without travelling far from where they reside.

These people, as well as some of those who do have local opportunities, tour more or less exclusively to engage in their NCA pastime. By contrast other NCA tourists travel for several reasons, only one of which is to spend some time, say, bird watching, river kayaking, collecting sea glass, or hiking a trail or two.

Regardless of which of these two approaches is taken to pursuing an NCA while touring, it is clear that such activity generates a strong desire in many participants to do it in different places away from the familiar ones near home. Still there seem to be people who are primarily enamoured of the nature they find close to home and little else. Thus, they are mainly interested in fishing local lakes and streams, photographing or painting local birds or animals, gathering mushrooms in nearby fields, or riding on horseback over trails in their area. In short the special interest tourism industry can profit from catering to the NCAs with large followings such as bird watching, certain kinds of hunting and fishing, snowmobiling and camping trips on horseback. But their potential clientele will never consist of the entire list of all people who go in for one of these activities.

Consumption and the NCAs

Stebbins (2009b) has recently set out a set of conceptual distinctions that can help us understand the place of consumption in NCA. He starts by observing that a substantial amount of consumption today has little or nothing to do with leisure, exemplified in buying toothpaste, life insurance, accounting services, natural gas for home heating, transit tickets for getting to work and so forth. Such consumption, call it *obligatory consumption*, shows in itself that we cannot regard leisure (un-coerced activity) and consumption as the same.

In *leisure-based consumption*, it is critical to distinguish whether the leisure component of a particular activity is directly and solely dependent on acquiring a good or service (e.g., buying a CD, concert ticket or session of massage) or whether purchase of something is prerequisite to a set of conditions that, much more centrally, shapes the activity as a leisure experience. In other words, is consumption an initiator or facilitator of a leisure experience. In *initiatory, leisure-based consumption* someone buys, say, a ticket enabling transportation by boat to a scuba diving site, a fee for a pilot to fly a plane from which to sky dive or a license with which to hunt or fish. In such consumption the purchaser proceeds more or less directly to use the purchased item. Here leisure and consumption do seem to be inextricably intertwined – an identity.

All these examples are serious leisure NCAs, even though most initiatory, leisure-based consumption in the modern world appears to be in service of casual leisure interests.

Facilitative, leisure-based consumption is different. Here the acquired item only sets in motion a set of activities, which when completed, enable the purchaser to use the item in a satisfying or fulfilling leisure experience. This is hardly the alienating consumption against which the critics of mass society raged (e.g., Marcuse, 1964; Lefebvre, 1991). As an example note that amateur botanists, if they are to operate at all as scientists, must first either rent or purchase a microscope – a consumptive act. Yet their most profound leisure experience is competently gathering specimens of plants from swamps or forests and then, with the acquired microscope (a consumer product), examining them for scientifically useful information. Moreover this profound leisure experience might be further facilitated by subscribing to a professional journal and buying gas to travel to the field in search of specimens.

The following excerpt from an article by Wade Bourne, written for Ducks Unlimited, addresses this issue in duck hunting:

The minimalists

By sticking to the basics, these waterfowlers have found success without spending a fortune on gear and gadgets.

Many duck hunters go for the gusto in pursuit of their sport. They build blinds that could serve as second homes. They acquire every motorized decoy available. They purchase expensive boats, guns, calls, and other gear. For this group, the sky is the limit, and the season – or the preparation for it – never ends.

Then there are the minimalists. These are hunters at the other extreme. Their boats and gear are ordinary. Their decoy spreads are small and without gimmicks. Their blinds are basic. And their mindset is to keep their hunting strategy as simple as possible. Find the birds, toss out just enough decoys, hide well, and load up.

Which approach is better? Each hunter must answer this question individually. Surely many prefer the all-out approach. Their energy and resources for this sport match their passion for it.

But just because the minimalists don't go for the grandiose doesn't mean they are any less zealous about duck hunting. Indeed, the opposite is often true. Many hunters take pride in keeping things simple. Plus, doing so allows them to move with the birds without encumbrance from "stuff." Here's a look at how four hunters pursue this back-to-basics style of duck hunting.

(http://www.ducks.org/Hunting/HuntingTips/3664/
TheMinimalists.html?poe=huntingtips,
retrieved 11 April 2010)

In these examples one or more consumer purchases or rentals are necessary steps to experiencing the cherished leisure. Still leisure activities exist for which no facilitative consumption whatsoever is needed to participate in them, where consumption and leisure are clearly separate spheres. This is *non-consumptive leisure* (described in Stebbins, 2009b, pp. 118–126). Such pursuits abound in casual and serious leisure, with non-consumptive projects being possible as well. In non-consumptive leisure, activities cost nothing, or at most, require only relatively small amounts of money.

It is possible that a survey of NCAs in a modern society would reveal that they offer the greatest number of opportunities for non-consumptive leisure. Thus, it has been argued that collectors of natural objects are likely to find non-consumptive leisure (Stebbins, 2009b, Chapter 5). Of course this person must usually consume some petrol to reach the forest to collect insects or fossils, the mountains to collect rocks or minerals or the shore to gather shells or sea life. Furthermore some of these enthusiasts may conclude that going to these places is difficult enough to justify owning a four-wheel drive vehicle, which for them removes this hobby from the category of non-consumptive leisure. And some rock hounds may want, among other equipment, a microscope to examine more closely what they have found. Collectors of natural objects of any kind, to ensure decent participation in their activities, may also need to buy special clothing. Day hikers and open water swimmers offer additional examples of largely non-consumptive leisure.

Much of what has been said in this section can be summarized in three generalizations. One, consumption in relation to nature challenge leisure, to the extent the first is either initiatory or facilitative of the second, is, in substantial part, a practical process – to enable pursuit of a leisure activity. According to the requirements of the activity, the participant will have to buy a particular good or service. Conspicuous consumption is an exception to this generalization, for in the world of NCAs as elsewhere it is certainly possible to show off with enviable goods or services priced substantially above what is needed for the consumer to engage in the activity at his or her level of accomplishment.

A second generalization follows, namely, that for the participant the essence of any consumption-based leisure experience lies in actually doing the core activity or activities. Moreover this essence differs sharply from the practical experiences of buying related goods and services. Thus, no matter how fine and expensive the recently purchased trumpet, running shoes or set of chisels, the unforgettable experience of playing a concerto, participating in a race or making a chair is what keeps amateurs and hobbyists in these activities coming back for more. Third, for some kinds of leisure, such monetary outlays are largely if not wholly unnecessary; this is the realm of non-consumptive leisure.

Finally, for some people, participation in an NCA is intertwined with the idea of a life that is 'unencumbered' by material goods. This was the case, for example, with the mountaineers in Davidson's (2006, pp. 139–140) study where the virtue of non-materialism was seen as springing from the simplicity and self-reliance of experiences in the mountains which they felt 'keeps it all in perspective' and makes them realize that they 'don't need so much stuff'.

Sustainability and the NCAs

In the concluding sections of Chapters 2 through 7, we examined how the NCAs covered in each chapter related to sustainability. Now we must consider the complex picture that emerges when we pool these observations. That picture is consistent with the conclusion of Fresque and Plummer (2009) in their review of the relevant literature, namely, that outdoor activities can unfavourably affect in many ways the environment they are pursued in. We examine these diverse effects within the framework of the seven principles for using the natural environment in the most sustainable way, principles promulgated by the Leave No Trace Center for Outdoor Ethics:

Leave no trace seven principles
1. Plan ahead and prepare
2. Travel and camp on durable surfaces
3. Dispose of waste properly
4. Leave what you find
5. Minimize campfire impacts
6. Respect wildlife
7. Be considerate of other visitors

(This copyrighted information has been reprinted with permission from Leave no Trace Center for Outdoor Ethics www.LNT.org, retrieved 14 April 2010.)

Let us look at each principle in detail. First, failure to plan ahead properly may force emergency misuse of the environment. For instance, a camper who has forgotten to pack a shovel must defecate on the ground instead of digging a hole, which is the proper way to dispose of human waste in the wilds (principle 3). Walking and camping on durable surfaces (those already packed from use) are more environmentally friendly than walking or camping on unused surfaces. Principle 4 urges activity participants to avoid picking flowers, leaves and berries, for example, and to avoid removing artefacts created and placed by past and present native peoples. The impact of campfires is reduced by using the same sites that others have before and, in the case of new sites, by constructing them with minimal change to the physical surroundings. Better yet, if feasible, bring a propane stove. Respecting wildlife means not feeding birds and animals, not disturbing their habitats and not harassing them (e.g., some snowmobilers chase coyotes, some off-leash dogs chase deer), to list a few such practices. Finally, being considerate of other users of the wilderness reduces their need to compensate at times for inconsiderate behaviour by damaging the environment. One common example in hiking occurs when large groups congregate on a trail, which forces approaching users who want to skirt the gathering to walk off-trail.

These principles apply, albeit differently, to all the NCAs discussed in this book. The land-based activities are, in general, the most likely to violate them. But those pursued in the air and on or in the water can also be threatening, in that to do them, participants must start and finish on land. So wilderness canoeists can damage landing sites as they start and end portages, and pilots of hang gliders may have to walk over virgin terrain once they have landed. Using cross-country ski trails groomed and set on snow away from summer-time roads or hiking trails (durable surfaces) can, through accumulative pressure, eventually damage the vegetation underneath. Using cars and trucks to gain access to starting points for activities undertaken in water or air and on ice or snow holds considerable potential for maltreating the environment when access to those points and parking space near them are inadequate. A hunter once described to one of the authors how, with a power saw, he cut approximately one hundred meters of crude road through scrub spruce, so as to more easily retrieve with his truck the moose he had just shot.

In fact, except for volunteer NCAs, none in itself fosters environmental improvement. Instead challenging nature leads only to a degree of impact on the NCA's environment such that the impact is (1) none or slight, (2) moderate or (3) considerable. Moreover the proportion of NCAs falling into the none-slight, moderate and considerable categories in each element varies from element to element. Let us look at these patterns for each element.

The activities in the air are, as a group, the least damaging. A main problem, as just noted, is coming and going on land to start and complete the activity. Moreover some aircraft disturb birds. The ice- and snow-based hobbies are the next most environmentally friendly. Here, however, in addition to possible pre- and post- activity harmful to the land, the chance of pollution from garbage left by users after eating or winter camping also exists. Next most threatening are the water-based hobbies. With them there is possible pre- and post-activity damage as well as pollution from gasoline engines and waste left in the river or lake or along shorelines having been pitched out from a boat.

The land-based activities are in general, as already noted, the most likely to destroy nature's integrity. That said we have yet to examine the environmental impact of the floral and faunal NCAs, none of which is essentially a trail-based activity. Trails are often used to reach to that part of a forest or field where the flora and fauna of interest live. But after that to pick, photograph, collect, fish or hunt it is commonly necessary to walk off-trail, sometimes for quite a distance. In this manner the ground and vegetation passed over are damaged, albeit only temporarily in most instances. Few environments are so sensitive that one or two people tramping over them once a year will cause irreparable harm. And the picking of leaves and mushrooms and the collecting of rocks and insects is likewise so minuscule that they may still be regarded as sustainable activities. Note, too, that motorized NCAs on land and in water and air commensurate with their prevalence, pollute with their carbon emissions and levels of noise.

Championing sustainability

In part because nature is awe-inspiring for nature challenge enthusiasts and in part because they want it to remain as pristine as possible for them, they often make fine champions of sustainability. For after all damaging an environment that they hold in awe works against this cherished value. Meanwhile the love for their NCA is also substantially diluted. At least this is how it works in principle.

Nevertheless, in practice, the picture is much more complicated. It seems that wonderment of nature is a variable, not a nominal classification of yes I have it or no I do not. At one end of the continuum we find the people profoundly in awe of the nature they behold. At the other end are those who 'couldn't care less'; that they are in a natural environment seems to matter little to them. Probably most people regarding nature fall somewhere between these extremes. Now it seems reasonable to conclude that those who are truly in awe would be the least likely to want to modify what they are perceiving – leave no garbage, cut down no trees or shrubs, pick no flowers and so on. At the opposite end, since the environment represents nothing special for them, are the people who do leave garbage or pick flowers unless persuaded otherwise by signs prohibiting acts of this sort.

It appears that some NCAs attract, or quickly convert, people to the sustainability pole of this continuum. Cross-country skiers, backpackers, nature photographers and canoeists and kayakers number among the examples here. At the other pole we find, for instance, drivers of off-road motorcycles and four-wheel drive vehicles, snowmobilers, fishers and hunters (e.g., see Barringer & Yardley, 2007, on the controversial use of all-terrain vehicles). At both poles there are exceptions; for example, backpackers who fail to pack out their garbage when they leave a camp site and dirt bikers who militate among fellow enthusiasts against riding off-trail (which tears up soil and vegetation). Some empirical support for this proposition comes from research by, for example, Teisl and O'Brien (2003) and Theodori et al. (1998), who conclude that participation in outdoor recreation is positively associated with concern for and favourable behaviour towards the environment. Nonetheless evidence on this issue is still weak. Nevertheless there has been a long-standing link between NCA participants and advocacy for environmental issues, from early efforts to protect wilderness areas from development and commercial exploitation through to more recent movements such as the surf groups discussed in Chapter 3.

A selective examination of the websites of clubs organized to promote these activities is revealing. Thus, the Ontario Federation of Trail Riders posts on its website the following code of conduct for cyclists (only those elements bearing on the present discussion are included in this excerpt):

Code of Conduct

- Do Not Trespass on private property.
- Ride on existing trails.

- Respect nature.
- Comply with all legislation, bylaws and insurance requirements.
- Whatever you pack in, pack out. Do not litter, and leave the place better than you found it.
- Use trails only according to the permitted uses indicated. Some trails are seasonal and can experience problems in the spring.
- Check the trail conditions.

> (http://www.oftr.ca/about.php,
> retrieved 10 April 2010)

The International Mountain Bicycling Association in its six-point Rules of the Trail stresses the following:

2. Leave No Trace

Be sensitive to the dirt beneath you. Wet and muddy trails are more vulnerable to damage than dry ones. When the trail is soft, consider other riding options. This also means staying on existing trails and not creating new ones. Don't cut switchbacks. Be sure to pack out at least as much as you pack in.

> (http://www.imba.com/about/trail_rules.html,
> retrieved 10 April 2010)

The Ontario Federation of Snowmobile Clubs says that:

today, responsible riding is more important than ever, so please continue to:

- leave tracks not trash;
- maintain you sled;
- protect wildlife;
- keep it quiet;
- stay on the trail;
- respect sensitive areas;
- embrace new technologies;
- spread the word.

> (http://www.ofsc.on.ca/Trails/Enviro.asp,
> retrieved 10 April 2010)

In one of the forums sponsored by MichiganSportsman.com, readers can find expressions in a number of the messages of land owner antipathy towards hunters who leave garbage on the property of

the former (http://www.michiganforums.com/forum/showthread.php?
t=38378, retrieved 11 April 2010). Further examples of guidelines for
pilots, rowers and waterskiers were given in Chapters 2 and 3.

On the other hand, such warnings are rare, vague or non-existent in
activities like hiking, cross-country skiing, backcountry camping, and
kayaking and canoeing. A vague warning is a general exhortation to
engage in conservation or good practices in doing the activity. The good
practices include respecting the people encountered while pursuing the
activity and resisting the temptation to feed wildlife. Nevertheless the
latter may be conceived of as a sustainability issue, in that, birds and
animals who become habituated to human food are thereby put at risk.
In other words they can become dependent on such a food supply (and
thus at risk when it is unavailable) and, in the case of predatory animals, aggressive to the point of attacking people and their belongings to
acquire it.

Conclusions

Gary Fine (1997) argues that 'nature' is a cultural construction and a
contested one at that. We, at least in the West, make cultural choices,
electing to protect nature, view it as an organic entity or regard it
humanistically, if not philosophically. Furthermore some people see certain aspects of nature as exploitable resources. Others look on it in more
than one of these ways and possibly others yet to be identified. The con-
tents of this book are commensurate with Fine's proposition: hobbyists
and amateurs pursuing NCAs hold the nature they encounter in awe
(the humanistic choice). We have argued that they also want to protect
it, even while the tendency to do so varies among the NCAs.

We may add to Fine's catalogue of constructions of nature another
main theme of this book, namely, that nature, or more precisely a
particular feature of it, offers to some serious leisure participants a challenge. Thus, the void flowing away from a mountain cliff is more than
something beautiful, views from which should remain undefiled by
humankind. It also presents a challenge for the pilot of a hang glider to
float gracefully through its space to the ground below. For fishers their
quarry is not only fascinating to watch in its natural habitat, it is also
a challenge to hook and land. As a final illustration the frozen waterfall
is magnificent in its shape and colour, but to ice climbers it offers the
additional attraction of getting to its summit.

While we have included indicative data when it could be found,
we lack decent data on the extent of participation in NCAs, a not

unexpected situation since the concept is new and a survey instrument capable of measuring serious leisure, NCAs included is of only recent origin (Gould et al., 2008). Fresque and Plummer (2009) conclude from their review of the literature that participation in outdoor recreational activities is increasing in both Canada and the United States. A report on the American scene produced by the Outdoor Foundation (2010) gives a more nuanced picture of an increase in overall participation, but a decrease in more costly activities (because of the economic situation). Meanwhile Pergams and Zaradic (2008) have data for the United States showing a general decrease. Since NCAs make up only a proportion of all outdoor recreation, the jury is still out as to whether interest in them in particular is rising, falling or holding steady. The Serious Leisure Inventory Measure developed by Gould and colleagues should enable us to eventually answer this question.

Notes

1 Nature Challenge Activities

1. One might question listing groomed alpine skiing and snowboarding here. Our justification for doing so is that large swathes of forest are typically cut down to develop the runs, resulting thus in a major modification of the environment. At times this environment is further modified by adding fences to demarcate the runs and making snow to give them the depth that nature has failed to provide. Lifts lining the sides of the runs also modify nature. The same may be said for the development of golf courses.
2. Mineralogy could be added to this list, but space limitations prevent covering it in this book.
3. Why did we not use 'awesome' as one of our adjectives to qualify 'awe'? Our answer is that awesome is widely used in popular language as a synonym to describe objects, experiences and situations experienced as terrific, or great. This usage is too broad for our purposes, for our interest lies specifically in the feeling of wonder. Note as well another main sense of 'awe', as denoting terror. If something goes terribly wrong during an NCA, terrifying awe may be the emotional response, which is not, however, what participants seek.
4. The serious leisure perspective and its three forms are discussed in considerably greater detail in Stebbins (1992, 2001a, 2007).
5. Stebbins has recently taken to using the term 'fulfilment', because it points to a fulfilling experience, or more precisely, to a set of chronological experiences leading to development to the fullest of a person's gifts and character, to development of that person's full potential, which is certainly both a reward and a benefit of serious leisure. 'Satisfaction', the term Stebbins once used, sometimes refers to a satisfying experience that is fun or enjoyable (also referred to as gratifying). In another sense this noun may refer to meeting or satisfying a need or want. In neither instance does satisfaction denote the preferred sense of fulfilment just presented (Stebbins, 2004b).
6. The Hash House Harrier treasure hunt has a serious leisure counterpart in 'geocaching', which in turn is related to orienteering (see Chapter 4).
7. For a partial review of the scientific literature on risk in leisure, see Stebbins (2005a, pp. 13–14, 21).
8. Georg Simmel (1971) wrote the classic treatise on the adventure, though he conceptualized it in very general terms. By contrast Johnson's conceptualization focuses narrowly on challenging activities. The latter is therefore better suited to this book.

3 Water

1. Participation is measured by the Outdoor Foundation as persons over the age of six who participated at least once in the previous 12 months. Sport England

measures more frequent participation for the adult population (aged 16 and over). Once a month participation equates to at least one day in the previous 28 days (see www.sportengland.org/research/active_people_survey.aspx. for more information).

2. Participation is measured by the Australian Sports Commission as persons over the age of 15 who participated at least once in the previous 12 months.
3. Nicholas Hayes, in his book *Saving Sailing*, also notes a decline in participation of more than 40 per cent since 1997 and 70 per cent since 1979 (http://www.savingsailing.com/Home/Book.html, retrieved 19 May 2010).
4. Almost 8 million people, or 2.8 per cent of the population went jet skiing in the United States in 2008 (Outdoor Foundation, 2009a).

4   Land

1. A 'pitch' is a rope length and the numbers used here refer to the numerical grading system for rock climbs used in North America.
2. For further discussion of the differences between adventure and sport climbing, see Stebbins (2005a), Lewis (2004), Kiewa (2002), Donnelly (2003) and Heywood (1994).

5   Flora and Fauna

1. Kite and pole aerial photography, which *is* primarily a hobby, is nevertheless excluded from this book on grounds that, as with kite flying, in general, the participant is not directly in nature (see Chapter 1).
2. Whether some collectors are in fact amateurs rather than hobbyists will have to be decided by research on the presence or absence of professionalism among full-time curators of various kinds of collections.

Bibliography

Acton, J. (2002). *The Man Who Touched the Sky*. London: Hodder and Stoughton.

Addison, G. (2000). *Whitewater Rafting*. London: New Holland.

Ainley, M. G. (1980). The contribution of the amateur to North American ornithology: A historical perspective. *Living Bird*, 18, 161–177.

Armstrong, B. (2005). *Getting Started in Powerboating*, 3rd ed. Camden, ME: International Marine/McGraw-Hill.

Arnold, D. (2006). Snowkiting takes off. *The Boston Globe* (online edition), 19 February.

Arnould, E. J., & Price, L. L. (1993). River magic: Extraordinary experience and the extended service encounter. *Journal of Consumer Research*, 20, 24–45.

Australian Sports Commission (2008). *Participation in Exercise, Recreation and Sport Survey 2008 Annual Report*. Retrieved 1 June 2010 from http://www.ausport.gov.au/__data/assets/pdf_file/0004/304384/ERASS_Report_2008.pdf.

Aversa, A. (1986). Notes on entry routes into a sport/recreational role: The case of sailing. *Journal of Sport and Social Issues*, 10, 49–59.

Bain, H. (2001). Light airs. *The Dominion Post*. 27 October.

Barringer, F., & Yardley, W. (2007). Surge in off-roading stirs dust and debate in West. *New York Times*, 30 December, online edition.

Beaudonnat, E. (2006). *Kiteboarding Vision: The Essential To Discover, Learn and Improve Kiteboarding*. Puerto Plata: International Kiteboarding Organisation.

Bell, M. (2003). 'Another kind of life': Adventure racing and epic expeditions. In R. E. Rinehart & S. Sydnor (Eds.), *To the Extreme: Alternative Sports, Inside and Out* (pp. 219–253). Albany, NY: State University of New York Press.

Bennett, J. (1995). *The Complete Motorcycle Book: A Consumer's Guide*. New York: Facts on File.

Boese, K., & Spreckels, C. (2008). *Kitesurfing: The Complete Guide* (J. Roberts, Trans.). Chichester, West Sussex, England; Hoboken, NJ: Wiley.

Booth, D. (1996). Surfing. In D. Levinson & K. Christensen (Eds.), *Encyclopedia of World Sport: From Ancient Times to the Present* (pp. 984–989). Santa Barbara, CA: ABC-CLIO.

Booth, D. (2004). Surfing: From one (cultural) extreme to another. In B. Wheaton (Ed.), *Understanding Lifestyle Sports: Consumption, Identity and Difference* (pp. 94–109). Oxon; New York: Routledge.

Borrie, W., & Roggenbuck, J. W. (2001). The dynamic, emergent, and multi-phasic nature of on-site wilderness experiences. *Journal of Leisure Research*, 33, 202–228.

Boyce, J. (2004). *The Ultimate Book of Power Kiting and Kiteboarding*. Guilford, Conn.: Lyons Press.

Boyle, R. H. (1959). An absence of wood nymphs. *Sports Illustrated*, 11 (11), E5–E8.

Bridgers, L. (2003). Out of the gene pool and into the food chain. In R. E. Rinehart & S. Sydnor (Eds.), *To the Extreme: Alternative Sports, Inside and Out* (pp. 179–189). Albany, NY: State University of New York Press.

Bryan, H. (1979). *Conflict in the Great Outdoors: Toward an Understanding and Managing of Diverse Sportsmen Preferences*. Tuscaloosa, AL: University of Alabama Press.

Burton, J. (2003). Snowboarding: The essence is fun. In R. E. Rinehart & S. Sydnor (Eds.), *To the Extreme: Alternative Sports, Inside and Out* (pp. 401–406). Albany, NY: State University of New York Press.

Butcher, S., & Sassi, E. (2007). Dogsledding. *Microsoft Encarta Online Encyclopedia 2007*. Retrieved 29 November 2009 from http://encarta.msn.com.

Cant, S. G. (2003). 'The tug of danger with the magnetism of mystery': Descents into 'the comprehensive, poetic-sensuous appeal of caves'. *Tourist Studies*, 3 (1), 67–81.

Carey, K. (1994). *A Beginner's Guide to Airsports*. London: A & C Black.

Celsi, R. L., Rose, R. L., & Leigh, T. W. (1993). An exploration of high-risk consumption through skydiving. *Journal of Consumer Research*, 20 (1), 1–23.

Choate, J. E. (1957). Recreational boating: The nation's family sport. *The ANNALS of the American Academy of Political and Social Science* (313), 109–112.

Cnaan, R. A., Handy, F., & Wadsworth, M. (1996). Defining who is a volunteer: Conceptual and empirical considerations. *Nonprofit and Voluntary Sector Quarterly*, 25, 364–383.

Coakley, J. (2001). *Sport in Society: Issues and Controversies*, 7th ed. New York: McGraw-Hill.

Cockroft, A. (1997). *4WD North Island: 80 Off Road Adventures*. Christchurch: Shoal Bay Press.

Coltman, D. W., O'Donoghue, P., Jorgenson, J. T., Hogg, J. T., Strodbeck, C., & Festa-Bianchet, M. (2003). Undesirable evolutionary consequences of trophy hunting. *Nature*, 426 (6967), 655–658.

Cotter, J. (2003). Eco (ego?) challenge: British Columbia, 1996. In R. E. Rinehart & S. Sydnor (Eds.), *To the Extreme: Alternative Sports, Inside and Out* (pp. 207–217). Albany, NY: State University of New York Press.

Crawford, S. J. (1996). Parachuting. In D. Levinson & K. Christensen (Eds.), *Encyclopedia of World Sport: From Ancient Times to the Present* (pp. 719–722). Santa Barbara, CA: ABC-CLIO.

Crawford, T. (2008). Scarier signs touted to keep skiers in bounds. *The Calgary Herald*, Saturday, 5 January, p. A12.

Crooks, P. (2008). Walnut Creek amateur botanist discovers two new species of plants! *Diablo: The Magazine of the San Francisco East Bay* (online edition), June 2008.

Crystal, D. (1994). *The Cambridge Encyclopedia*, 2nd ed. Cambridge, UK: Cambridge University Press.

Csikszentmihalyi, M. (1990). *Flow: The Psychology of Optimal Experience*. New York, NY: Harper & Row.

Cutlip, K. (1999). An eye on the sea. *Weatherwise*, 52 (1), 11–12.

Dant, T. (1998). Playing with things: Objects and subjects in windsurfing. *Journal of Material Culture*, 3 (1), 77–95.

Davidson, L. (2002). The 'Spirit of the Hills': Mountaineering in Northwest Otago, New Zealand 1882–1940. *Tourism Geographies*, 4 (1), 44–61.

Davidson, L. (2006). *A Mountain Feeling: The Narrative Construction of Meaning and Self Through a Commitment to Mountaineering in Aotearoa/New Zealand*.

Unpublished PhD, Monash University, Melbourne (Also available at http://researcharchive.vuw.ac.nz/handle/10063/803?show=full).

Davidson, L. (2008). Travelling light in hostile country: Mountaineering, commitment and the leisure lifestyle. In J. Caudwell, S. Redhead & A. Tomlinson (Eds.), *Relocating the Leisure Society: Media, Consumption and Spaces* (Vol. 101, pp. 77–95). Eastbourne, UK: Leisure Studies Association.

Dean, P. L. (1998). *Open Water Swimming: A Complete Guide for Distance Swimmers and Triathletes*. Champaign, IL: Human Kinetics.

Desmond, M. (2007). *On the Fireline: Living and Dying with Wildland Firefighters*. Chicago, IL: University of Chicago Press.

Dimmock, K. (2007). Scuba diving, snorkeling, and free diving. In G. Jennings (Ed.), *Water-Based Tourism, Sport, Leisure, and Recreation Experiences* (pp. 128–147). Oxford; New York: Elsevier.

Dimmock, K. (2009). Finding comfort in adventure: Experiences of recreational SCUBA divers. *Leisure Studies*, 28 (3), 279–295.

Donnelly, P. (2003). The great divide: Sport climbing vs. adventure climbing. In R. E. Rinehart & S. Sydnor (Eds.), *To the Extreme: Alternative Sports, Inside and Out* (pp. 291–304). Albany, NY: State University of New York Press.

Donnelly, P. (2004). Sport and risk culture. In K. E. Young (Ed.), *Sporting Bodies, Damaged Selves: Sociological Studies of Sports-Related Injury*. Oxford, United Kingdom: Elsevier.

Eassom, S. (2003). Mountain biking madness. In R. E. Rinehart & S. Sydnor (Eds.), *To the Extreme: Alternative Sports, Inside and Out* (pp. 191–203). Albany, NY: State University of New York Press.

Economist (The) (2001). Régine Cavagnoud. 17 November, 82.

Economist (The) (2005). Up off the couch. 22 October, 35.

Economist (The) (2008). Steve Fossett. 23 February, 106.

Ecott, T. (2001). *Neutral Buoyancy: Adventures in a Liquid World*. London: Michael Joseph.

Edmonds, W. A., Tenenbaum, G., Kamata, A., & Johnson, M. B. (2009). The role of collective efficacy in adventure racing teams. *Small Group Research*, 40 (2), 163–180.

Elliot, N. (2004). 'Soulriding' – The spirituality of snowboarding. Paper presented at the Annual Meeting of the British Association for the Sociology of Religion, Oxford, September.

Evans, J., Heikell, R., Jeffery, T., & O'Grady, A. (2007). *Sailing*. London: Dorling Kindersley.

Evers, C. (2006). How to surf. *Journal of Sport and Social Issues*, 30 (3), 229–243.

Farrell, E. (2006). *One Breath: A Reflection on Freediving*. Hatherleigh, Devon: Pynto.

Farrell, S. P. (2009). The urban deerslayer. *New York Times*, Wednesday, 25 November, online edition.

Favret, B., & Benzel, D. (1997). *Complete Guide to Water Skiing*. Champaign, IL: Human Kinetics.

Ferrell, J., Milovanovich, D., & Lyng, S. (2001). Edgework, media practices, and the elongation of meaning: A theoretical ethnography of the Bridge Day event. *Theoretical Criminology*, 5 (2), 177–202.

Fine, G. A. (1988). Dying for a laugh. *Western Folklore*, 47, 77–194.

Fine, G. A. (1997). Naturework and the taming of the wild: The nature of 'overpick:' in the culture of mushroomers. *Social Problems*, 44, 68–88.

Fine, G. A. (1998). *Morel Tales: The Culture of Mushrooming*. Cambridge, MA: Harvard University Press.

Fine, G. A. (2007). *Authors of the Storm: Meteorologists and the Culture of Prediction*. Chicago, London: University of Chicago Press.

Finnigan, C. (2001). *Microlighting: Affordable Aviation*. Crowood: Marlborough.

Fleury, R. (2007). Deeper blue. *BBC Focus*, 179, 30–36.

Floro, G. K. (1978). What to look for in a study of the volunteer in the work world. In R. P. Wolensky, & E. J. Miller (Ed.), *The small City and Regional Community* (pp. 194–202). Stevens Point, WI: Foundation Press.

Gedzelman, S. D. (1989). Cloud classification before Luke Howard. *Bulletin of the American Meteorological Society*, 70 (4), 381–395.

Georgeson, D., & Wilson, A. (2003). *The Leading Edge: A life in Gliding*. Christchurch: Shoal Bay Press.

Gibson, H., Willming, C., & Holdnak, A. (2002). We're gators…not just Gator fans: Serious leisure and University of Florida football. *Journal of Leisure Research*, 34, 397–425.

Glaser, B. G. (1978). *Theoretical Sensitivity: Advances in the Methodology of Grounded Theory*. Mill Valley, CA: Sociology Press.

Glaser, B. G., & Strauss, A. L. (1967). *The Discovery of Grounded Theory: Strategies for Qualitative Research*. Chicago, IL: Aldine Atherton.

Gluckman, R. (2008, 2 June). I work to kite. *Forbes Asia*, 4 (10), 70–72.

Goffman, E. (1961). *Asylums: Essays on the Social Situation of Mental Patients and Other Inmates*. Chicago, IL: Aldine.

Goffman, E. (1963). *Stigma: Notes on the Management of Spoiled Identity*. Englewood Cliffs, NJ: Prentice Hall.

Gould, J., Moore, D., McGuire, F., & Stebbins, R. A. (2008). Development of the serious leisure inventory measure. *Journal of Leisure Research*, 40, 47–68.

Goyer, N. (2004). *Air Sports: The Complete Guide to Aviation Adventure*. New York: McGraw-Hill.

Graver, D. K. (1999). *Scuba Diving*. Champaign, IL: Human Kinetics.

Hamblyn, R. (2001). *The Invention of Clouds: How an Amateur Meteorologist Forged the Language of the Skies*. New York: Farrar, Straus and Giroux.

Harper, M. (2007). *The Ways of the Bushwalker: On Foot in Australia*. Sydney: University of New South Wales Press.

Harrison, D. (1993). *Sea kayaking Basics*. New York: Hearst Marine Books.

Herfindal, I., Linnell, J. D. C., Moa, P. F., Odden, J., Austmo, L. B., & Andersen, R. (2005). Does recreational hunting of lynx reduce depredation losses of domestic sheep? *Journal of Wildlife Management*, 69, 1034–1042.

Heywood, I. (1994). Urgent dreams: Climbing, rationalization and ambivalence. *Leisure Studies*, 13 (3), 179–194.

Hudson, S., & Beedie, P. (2007). Kayaking. In G. Jennings (Ed.), *Water-Based Tourism, Sport, Leisure, and Recreation Experiences* (pp. 171–186). Oxford; New York: Elsevier.

Hummel, R. (2004). Hunting. In G. S. Cross (Ed.), *Encyclopedia of Recreation and Leisure in America*, vol. 1 (pp. 460–464). Detroit, MI: Thomson Gale.

Humphreys, D. (2003). Selling out snowboarding. In R. E. Rinehart & S. Sydnor (Eds.), *To the Extreme: Alternative Sports, Inside and Out* (pp. 407–428). Albany, NY: State University of New York Press.

Hunt, L. M., Haider, W., & Bottan, B. (2005). Accounting for varying setting preferences among moose hunters. *Leisure Sciences*, 27, 297–314.

Hurd, B. (2003). *Entering the Stone: On Caves and Feeling Through the Dark*. Boston: Houghton Mifflin.

Jennings, G. (2007a). Motorboating. In G. Jennings (Ed.), *Water-Based Tourism, Sport, Leisure, and Recreation Experiences* (pp. 46–63). Oxford; New York: Elsevier.

Jennings, G. (2007b). Sailing/cruising. In G. Jennings (Ed.), *Water-Based Tourism, Sport, Leisure, and Recreation Experiences* (pp. 23–45). Oxford; New York: Elsevier.

Johnson, R. (2003). Adventure. In J. M. Jenkins & J. J. Pigram (Eds.), *Encyclopedia of Leisure and Outdoor Recreation* (pp. 8–9). London: Routledge.

Jones, C. D., Hollenhorst, S., Perna, F., & Selin, S. (2000). Validation of the flow theory in an on-site whitewater kayaking setting. *Journal of Leisure Research, 32* (2), 247–261.

Jones, L. (2007). Whitewater rafting. In G. Jennings (Ed.), *Water-Based Tourism, Sport, Leisure, and Recreation Experiences* (pp. 153–170). Oxford; New York: Elsevier.

Kay, J., & Laberge, S. (2002). The 'new' corporate habitus in adventure racing. *International Review for the Sociology of Sport, 37* (1), 17–36.

Kay, J., & Laberge, S. (2004). 'Mandatory equipment': Women in adventure racing. In B. Wheaton (Ed.), *Understanding Lifestyle Sports: Consumption, Identity and Difference* (pp. 154–174). Oxon; New York: Routledge.

Keeney, E. B. (1992). *The Botanizers: Amateur Scientists in Nineteenth-Century America*. Chapel Hill, NC: University of North Carolina Press.

Kelly, J. R., & Warnick, R. B. (1999). *Recreation Trends and Markets: The 21st Century*. Champaign, IL: Sagamore.

Kiewa, J. (2002). Traditional climbing: Metaphor of resistance or metanarrative of oppression? *Leisure Studies, 21,* 145–161.

Koukouris, K. (2005). Beginners' perspectives of getting involved in orienteering in Greece. *Scientific Journal of Orienteering, 16,* 18–33.

Kuhne, C. (1998). *Canoeing*. Mechanicsburg, PA: Stackpole Books.

Langewiesche, W. (1998). *Inside the Sky: A Meditation on Flight*. New York: Pantheon.

Laurendeau, J., & Sharara, N. (2008). 'Women could be every bit as good as guys': Reproductive and resistant agency in two 'action' sports. *Journal of Sport and Social Issues, 32* (1), 24–47.

Lefevbre, H. (1991). *Critique of Everyday Life, vol. 1, Introduction,* trans. by J. Moore. London: Verso.

Lewis, N. (2004). Sustainable adventure: Embodied experiences and ecological practices within British climbing. In B. Wheaton (Ed.), *Understanding Lifestyle Sports: Consumption, Identity and Difference* (pp. 70–93). Oxon: Routledge.

Lipscombe, N. (1999). The relevance of the peak experience to continued skydiving participation: A qualitative approach to assessing motivations. *Leisure Studies, 18,* 267–288.

Littin, B. (1990). Citizen weather observers. *Weatherwise, 43* (5), 254–259.

Lois, J. (2003). *Heroic Efforts: The Emotional Culture of Search and Rescue Volunteers*. New York: New York University Press.

Lois, J. (2004). Gender and emotional management in the stages of edgework. In S. Lyng (Ed.), *Edgework: The Sociology of Risk-Taking* (pp. 117–152). Florence, KY: Routledge.

Loland, S., & Sandberg, P. (1995). Realizing ludic rationality in sport competitions. *International Review for the Sociology of Sport, 30* (2), 225–240.

Ludwig, R. P. (1996). Ballooning. In D. Levinson & K. Christensen (Eds.), *Encyclopedia of World Sport: From Ancient Times to the Present* (pp. 69–72). Santa Barbara, CA: ABC-CLIO.

Lunn, A. (1957). *A Century of Mountaineering 1857–1957*. London: Allen & Unwin.

Macaloon, J., & Csikszentmihalyi, M. (1983). Deep play and the flow experience in rock climbing. In J. C. Harris & R. J. Park (Eds.), *Play, Games and Sport in Cultural Contexts* (pp. 361–384). Champaign, IL: Human Kineticss.

MacFarlane, R. (2004). *Mountains of the Mind: Adventures in Reaching the Summit*. New York: Vintage Books.

MacMillan, D.C., & Leitch, K. (2008). Conservation with a gun: Understanding landowner attitudes to deer hunting in the Scottish highlands. *Human Ecology*, 36, 473–484.

Mahdavi, S. (2007). Amusements in Qajar Iran, *Iranian Studies*, 40 (4), 483–499.

Mann, D., & Schaad, K. (2001). *The Complete Guide to Adventure Racing: The Insider's Guide to the Greatest Sport on Earth*. New York: Hatherleigh Press.

Mann, M. J., & Leahy, J. E. (2009). Connections: Integrated meanings of ATV riding among club members in Maine. *Leisure Sciences*, 31, 384–396.

Marais, J. (2002). *Hiking: The Essential Guide to Equipment and Techniques*. London: New Holland.

Marais, J., & de Speville, L. (2004). *Adventure Racing*. Champaign, IL: Human Kinetics.

Mattos, B., & Middleton, A. (2004). *Advanced Kayaking & Canoeing: A Practical Guide to Paddling on White Water, Open Water and the Sea*. London: Southwater.

Marcuse, H. (1964). *One-Dimensional Man. Studies in the Ideology of Advanced Industrial Society*. Boston: Beacon Press.

McCairen, P. C. (1998). *Canyon Solitude: A Woman's Solo River Journey Through Grand Canyon*. Seattle, Wash: Seal Press.

Mckhann, M. (2001). Snowboarding. *Microsoft Encarta Encyclopedia Standard 2001*. Retrieved 23 January 2004 from http://encarta.com.

Meyersohn, R. (1970). The charismatic and the playful in outdoor recreation. *Annals of the American Academy of Political and Social Science*, 389, 35–45.

Microsoft Encarta Encyclopedia Standard 2001 (2007a.) Snowmobiling. Retrieved 19 August 2009 from http://encarta.com.

Microsoft Encarta Encyclopedia Standard 2001 (2007b). Ice Skating. Retrieved 16 August 2009 from http://encarta.com.

Microsoft Encarta Encyclopedia Standard 2001 (2007c). Ice Boating. Retrieved 22 August 2009 from http://encarta.com.

Miller, G. A. (2007). Trails and the hiking experience: A natural connection. *American Hiker*, Summer 2007, 4–7.

Miller, W. D. (1996). Soaring. In D. Levinson & K. Christensen (Eds.), *Encyclopedia of World Sport: From Ancient Times to the Present* (pp. 938–944). Santa Barbara, CA: ABC-CLIO.

Mullins, P. M. (2009). Living stories of the landscape: Perception of place through canoeing in Canada's north. *Tourism Geographies*, 11 (2), 233–255.

Olmsted, A.D. (1991). Collecting: Leisure, investment, or obsession? *Journal of Social Behavior and Personality*, 6, 287–306.

Osgood, W., & Hurley, L. (1975). *The Snowshoe Book*, 2nd ed, rev. Brattleboro, VT: The Stephen Greene Press.

Outdoor Foundation (2009a). *Outdoor Recreation Participation Report 2009.* Washington, DC: The Outdoor Foundation. Retrieved 26 May 2010 from http://www.outdoorfoundation.org/pdf/ResearchParticipation2009.pdf.

Outdoor Foundation (2009b). *A Special Report on Paddlesports 2009: Kayaking, Canoeing, Rafting.* Washington, DC: The Outdoor Foundation. Retrieved 30 May 2010 from http://www.outdoorfoundation.org/pdf/ResearchPaddlesports.pdf.

Outdoor Foundation (2010). *Outdoor Recreation Participation Top Line Report 2010.* Washington, DC: The Outdoor Foundation. Retrieved 30 May 2010 from http://www.outdoorfoundation.org/pdf/ResearchParticipation2010 Topline.pdf.

Paduda, J. (1992). *The Art of Sculling.* Camden, ME: Ragged Mountain Press.

Pagen, D. (1996). Hang gliding. In D. Levinson & K. Christensen (Eds.), *Encyclopedia of World Sport: From Ancient Times to the Present* (pp. 405–408). Santa Barbara, CA: ABC-CLIO.

Parr, A. R. (1996). Sandyachting. In D. Levinson & K. Christensen (Eds.), *Encyclopedia of World Sport: From Ancient Times to the Present* (pp. 865–868). Santa Barbara, CA: ABC-CLIO.

Pearson, K. (1982). Surfies and clubbies in Australia and New Zealand. *Journal of Sociology,* 18 (1), 5–15.

Pergams, O. R. W., & Zaradic, P. A. (2008). Evidence for a fundamental and pervasive shift away from nature-based recreation. *PNAS,* 105 (7), 2295–2300.

Piggott, D. (1997). *Gliding: A Handbook on Soaring Flight,* 7th ed. London: A & C Black.

Piggott, D. (2000). *Beginning Gliding.* London: A & C Black.

Pigram, J. J. (2003). Water-based recreation. In J. M. Jenkins & J. J. Pigram (Eds.), *Encyclopedia of Leisure and Outdoor Recreation* (pp. 543–547). London; New York: Routledge.

Ponting, J. (2008). *Consuming Nirvana: An Exploration of Surfing Tourist Space.* Unpublished PhD, University of Technology Sydney, Sydney.

Poole, E. (2003). Family mourns vibrant sisters. *Calgary Herald,* Wednesday 2 July, p. A1.

Preston-Whyte, R. (2002). Constructions of surfing space at Durban, South Africa. *Tourism Geographies,* 4 (3), 307–328.

Presque, J., & Plummer, R. (2009). Accounting for consumption related to outdoor recreation: An application of ecological footprint analysis. *Leisure/Loisir,* 33, 589–614.

Prettyman, B. (2006). Catch & cook: Hooked on wild picks. *Sporadic Press: Journal of the San Diego Mycological Society,* 11(1). Retrieved 08 January 2010 from http://www.sdmyco.org/200609Sept.htm.

Proctor, R. (2004). Rockhounding. In G. S. Cross (Ed.), *Encyclopedia of Recreation and Leisure in America,* vol. 2 (pp. 224–225). Detroit, MI: Thomson Gale.

Randall, M., Macbeth, J., & Newsome, D. (2006). Investigating the impacts of off-road vehicle activity in Broome, North-Western Australia: A preliminary appraisal. *Annals of Leisure Research,* 9 (1–2), 17–42.

Reid, E. (1991). The green hunter. *Pathways,* 3 (June), 11–15.

Redgrave, S. (1995). *Steven Redgrave's Complete Book of Rowing.* London: Partridge Press.

Reekie, S. H. M. (1996). Sailboarding. In D. Levinson & K. Christensen (Eds.), *Encyclopedia of World Sport: From Ancient Times to the Present* (pp. 847–849). Santa Barbara, CA: ABC-CLIO.

Richins, H. (2007). Motorized water sports. In G. Jennings (Ed.), *Water-Based Tourism, Sport, Leisure, and Recreation Experiences* (pp. 69–94). Oxford; New York: Elsevier.

Roberts, M. B. (2007). *Crew: The Rower's Handbook*. New York: Sterling Publishing.

Rogers, A. J. (1960). Mature science – retarded profession. *The Florida Entomologist*, 43, 155–162.

Ross, K. (2008). *Going Bush: New Zealanders and Nature in the Twentieth Century*. Auckland: Auckland University Press.

Ryan, C. (2007). Surfing and windsurfing. In G. Jennings (Ed.), *Water-Based Tourism, Sport, Leisure and Recreation Experiences* (pp. 95–111). Oxford; New York: Elsevier.

Sande, A. (2001). The natural social order. *Sosiologisk Tidsskrift*, 9 (4), 350–371.

Schlatter, B. E., & Hurd, A. R. (2005). Geocaching: 21st-century hide-and-seek. *Journal of Physical Education, Recreation & Dance*, 76 (7), 28–32.

Scott, D. (2004). Bird watching. In G. S. Cross (Ed.), *Encyclopedia of Recreation and Leisure in America*, vol. 2 (p. 103). Detroit, MI: Thomson Gale.

Simmel, G. (1971). *George Simmel on Individuality and Social Forms*, edited by D.N. Levine. Chicago, IL: University of Chicago Press.

Smith, A., & Wagner, M. (1998). *Ballooning*. Nr Yeovil, Somerset: Patrick Stephens.

Smith, D. H., Stebbins, R. A., & Dover, M. (2006). *A Dictionary of Nonprofit Terms and Concepts*. Bloomington, IN: Indiana University Press.

Spearpoint, G. (1985). *Walking to the Hills: Tramping in New Zealand*. Auckland: Reed Methuen.

Sport England (2009). *Active People Survey 3: Once a Month Sport Participation Rates by Sport*. Retrieved 25 May 2010 from http://www.sportengland.org/research/active_people_survey/idoc.ashx?docid=cc9d9bd3-d9cd-44d3-bdbc-fb35d4ac95fd&version=2.

Staseson, H., & Komarnicki, J. (2010). Rewards worth the risk, say adrenalin junkies. *Calgary Herald*, Monday, 15 March.

Stebbins, R. A. (1979). *Amateurs: On the Margin Between Work and Leisure*. Beverly Hills, CA: Sage Publications.

Stebbins, R. A. (1992). *Amateurs, Professionals and Serious Leisure*. Montreal, QC: McGill-Queen's University Press.

Stebbins, R. A. (1993). *Canadian Football: The View from the Helmet*. Toronto, ON: Canadian Scholars Press.

Stebbins, R. A. (1994). The liberal arts hobbies: A neglected subtype of serious leisure. *Loisir et société/Society and leisure*, 17, 173–186.

Stebbins, R. A. (1996). *The Barbershop Singer: Inside the Social World of a Musical Hobby*. Toronto, ON: University of Toronto Press.

Stebbins, R. A. (1997). Casual leisure: A conceptual statement. *Leisure Studies*, 16, 17–25.

Stebbins, R. A. (1998a). *After Work: The Search for an Optimal Leisure Lifestyle*. Calgary, AB: Detselig.

Stebbins, R. A. (1998b). *The Urban Francophone Volunteer: Searching for Personal Meaning and Community Growth in a Linguistic Minority*, Vol. 3 (2). Seattle, WA: Canadian Studies Center, University of Washington.

Stebbins, R. A. (2001a). *New Directions in the Theory and Research of Serious Leisure*. Lewiston, NY: Edwin Mellen.

Stebbins, R. A. (2001b). *Exploratory Research in the Social Sciences*. Thousand Oaks, CA: Sage.

Stebbins, R. A. (2001c). Volunteering – mainstream and marginal: Preserving the leisure experience. In M. G. M. Foley (Ed.), *Volunteering in Leisure: Marginal or Inclusive?* (Vol. 75, pp. 1–10). Eastbourne, UK: Leisure Studies Association.

Stebbins, R. A. (2002). *The Organizational Basis of Leisure Participation: A Motivational Exploration*. State College, PA: Venture Publishing.

Stebbins, R. A. (2003). Casual leisure. In J. M. Jenkins & J. J. Pigram (Eds.), *Encyclopedia of Leisure and Outdoor Recreation* (pp. 44–46). London: Routledge.

Stebbins, R. A. (2004a). *Between Work and Leisure: The Common Ground of Two Separate Worlds*. New Brunswick, NJ: Transaction Publishers.

Stebbins, R. A. (2004b). Fun, enjoyable, satisfying, fulfilling: Describing positive leisure experience. *Leisure Studies Association Newsletter*, 69 (November), 8–11 (also available at www.soci.ucalgary.ca/seriousleisure – Digital Library, 'Leisure Reflections No. 7').

Stebbins, R. A. (2004c). Pleasurable aerobic activity: A type of casual leisure with salubrious implications. *World Leisure Journal*, 46 (4), 55–58.

Stebbins, R. A. (2005a). *Challenging Mountain Nature: Risk, Motive, and Lifestyle in Three Hobbyist Sports*. Calgary, AB: Detselig.

Stebbins, R. A. (2005b). Choice and experiential definitions of leisure. *Leisure Sciences*, 27, 349–352.

Stebbins, R. A. (2005c). Project-based leisure: Theoretical neglect of a common use of free time. *Leisure Studies*, 24, 1–11.

Stebbins, R. A. (2007). *Serious Leisure: A perspective for Our Time*. New Brunswick, NJ: Transaction.

Stebbins, R. A. (2009a). New leisure and leisure customization. *World Leisure Journal*, 51 (2), 78–84.

Stebbins, R. A. (2009b). *Leisure and Consumption: Common Ground, Separate Worlds*. Houndmills, Basingstoke, UK: Palgrave Macmillan.

Stebbins, R. A. (2009c). Case study 7.1, Mountain climbing and serious leisure. In J. Higham & T. Hinch (Eds.), *Sport and Tourism: Globalisation, Mobility, and Identity* (pp. 131–133). Oxford, UK: Elsevier Butterworth Heinemann.

Stebbins, R. A. (2009d). *Personal Decisions in the Public Square: Beyond Problem Solving into a Positive Sociology*. New Brunswick, NJ: Transaction.

Stebbins, R. A. (2010). Flow in serious leisure: Nature and prevalence. *Leisure Studies Association Newsletter*, 87 (November), in press (also available at www.soci.ucalgary.ca/seriousleisure – Digital Library, 'Leisure Reflections No. 24').

Stranger, M. (1999). The aesthetics of risk: A study of surfing. *International Review for the Sociology of Sport*, 34 (3), 265–276.

Svenvold, M. (2006). *Big Weather: Chasing Tornados in the Heart of America*. New York: Henry Holt.

Sweet, B. (2006). *Powerboat Handling Illustrated: How to Make Your Boat Do Exactly What You Want It to Do*. Camden, Me.: International Marine/McGraw-Hill.

Taylor, M. R. (1996). *Cave passages: Roaming the Underground Wilderness*. New York: Scribner.

Teisl, M. F., & O'Brien, K. (2003). Who cares and who acts? Outdoor recreationists exhibit different levels of environmental concern and behavior. *Environment and Behavior*, 35, 506–522.

Temple, P. (1969). *The World at Their Feet*. Christchurch: Whitcombe & Tombs.

Theodori, G. L., Luloff, A. E., & Willits, F. K. (1998). The association of outdoor recreation and environmental concern: Reexamining the Dunlap-Heffernan thesis. *Rural Sociology*, 63, 94–108.

Times (The) (2009). Francis Rogallo: Invented the Rogallo wing. 29 September. Retrieved 11 April 2010 from http://www.timesonline.co.uk/tol/comment/obituaries/article6852741.ece.

Townes, J. (1996a). Canoeing and kayaking. In D. Levinson & K. Christensen (Eds.), *Encyclopedia of World Sport: From Ancient Times to the Present* (pp. 168–173). Santa Barbara, CA: ABC-CLIO.

Townes, J. (1996b). Motor boating. In D. Levinson & K. Christensen (Eds.), *Encyclopedia of World Sport: From Ancient Times to the Present* (pp. 646–653). Santa Barbara, CA: ABC-CLIO.

Townes, J. (1996c). Rafting. In D. Levinson & K. Christensen (Eds.), *Encyclopedia of World Sport: From Ancient Times to the Present* (pp. 790–791). Santa Barbara, CA: ABC-CLIO.

Townsend, C. (2009). The journey. *The Great Outdoors (August)*, 78–79.

Unruh, D. R. (1979). Characteristics and types of participation in social worlds. *Symbolic Interaction*, 2, 115–130.

Unruh, D. R. (1980). The nature of social worlds. *Pacific Sociological Review*, 23, 271–296.

Watling, R. (1998). The role of the amateur in mycology – what would we do without them! *Mycoscience*, 39 (4), 513–522.

Westbrook, A., & Westbrook, P. (1966). *Trail Horses and Trail Riding*. London: Thomas Yoseloff.

Wheaton, B. (2000). 'Just do it': Consumption, commitment, and identity in the windsurfing subculture. *Sociology of Sport Journal*, 17, 254–274.

Wynveen, C. J., Kyle, G. T., & Sutton, S. G. (2010). Place meanings ascribed to marine settings: The case of the Great Barrier Reef Marine Park. *Leisure Sciences*, 32, 270–287.

Yoder, D. G. (1997). A model for commodity intensive serious leisure. *Journal of Leisure Research*, 29, 407–429.

Yoder, D. G. (2004a). Fishing, freshwater. In G. S. Cross (Ed.), *Encyclopedia of Recreation and Leisure in America*, vol. 1 (pp. 345–348). Detroit, MI: Thomson Gale.

Yoder, D. G. (2004b). Fishing, saltwater/deep sea. In G. S. Cross (Ed.), *Encyclopedia of Recreation and Leisure in America*, vol. 1 (pp. 348–351). Detroit, MI: Thomson Gale.

Index

Note: NCAs = nature challenge activities